NATIONAL RESEARCH INITIATIVE

A VITAL COMPETITIVE GRANTS PROGRAM IN FOOD, FIBER, AND NATURAL-RESOURCES RESEARCH

COMMITTEE ON AN EVALUATION OF THE US DEPARTMENT OF AGRICULTURE NATIONAL RESEARCH INITIATIVE COMPETITIVE GRANTS PROGRAM

BOARD ON AGRICULTURE AND NATURAL RESOURCES

NATIONAL RESEARCH COUNCIL

NATIONAL ACADEMY PRESS
Washington, DC

NATIONAL ACADEMY PRESS · 2101 Constitution Avenue, NW · Washington, D.C. 20418

This material is based upon work supported by the US Department of Agriculture, Cooperative State Research, Education, and Extension Service under agreement 97-COOP-2-5045. Any opinions, findings, conclusions, or recommendations expressed in this publication are those of the authors and do not necessarily reflect the view of the organizations or agencies that provided support for this project.

ISBN 0-309-07083-X

Printed in the United States of America.

THE NATIONAL ACADEMIES

National Academy of Sciences
National Academy of Engineering
Institute of Medicine
National Research Council

The **National Academy of Sciences** is a private, nonprofit, self-perpetuating society of distinguished scholars engaged in scientific and engineering research, dedicated to the furtherance of science and technology and to their use for the general welfare. Upon the authority of the charter granted to it by the Congress in 1863, the Academy has a mandate that requires it to advise the federal government on scientific and technical matters. Dr. Bruce M. Alberts is president of the National Academy of Sciences.

The **National Academy of Engineering** was established in 1964, under the charter of the National Academy of Sciences, as a parallel organization of outstanding engineers. It is autonomous in its administration and in the selection of its members, sharing with the National Academy of Sciences the responsibility for advising the federal government. The National Academy of Engineering also sponsors engineering programs aimed at meeting national needs, encourages education and research, and recognizes the superior achievements of engineers. Dr. William A. Wulf is president of the National Academy of Engineering.

The **Institute of Medicine** was established in 1970 by the National Academy of Sciences to secure the services of eminent members of appropriate professions in the examination of policy matters pertaining to the health of the public. The Institute acts under the responsibility given to the National Academy of Sciences by its congressional charter to be an adviser to the federal government and, upon its own initiative, to identify issues of medical care, research, and education. Dr. Kenneth I. Shine is president of the Institute of Medicine.

The **National Research Council** was organized by the National Academy of Sciences in 1916 to associate the broad community of science and technology with the Academy's purposes of furthering knowledge and advising the federal government. Functioning in accordance with general policies determined by the Academy, the Council has become the principal operating agency of both the National Academy of Sciences and the National Academy of Engineering in providing services to the government, the public, and the scientific and engineering communities. The Council is administered jointly by both Academies and the Institute of Medicine. Dr. Bruce M. Alberts and Dr. William A. Wulf are chairman and vice chairman, respectively, of the National Research Council.

COMMITTEE ON AN EVALUATION OF THE US DEPARTMENT OF AGRICULTURE NATIONAL RESEARCH INITIATIVE COMPETITIVE GRANTS PROGRAM

THOMAS N. URBAN, *Chair,* Des Moines, Iowa, Pioneer Hi-Bred International, Inc., Retired
PETER J. BARRY, University of Illinois, Urbana
FRANCIS F. BUSTA, University of Minnesota, St. Paul
MARY-DELL CHILTON, Novartis Seeds, Inc., Research Triangle Park, North Carolina
DARYL E. CHUBIN, National Science Foundation, Arlington, Virginia
ROBERT J. COLLIER, University of Arizona, Tucson
NOEL T. KEEN, University of California, Riverside
MICHAEL R. LADISCH, Purdue University, West Lafayette, Indiana
G. PHILIP ROBERTSON, Michigan State University, Hickory Corners
RONALD R. SEDEROFF, North Carolina State University, Raleigh
WILLIAM W. SIMPKINS, Iowa State University, Ames
ROBERT E. SMITH, R E Smith Consulting, Inc., Newport, Vermont
FREDRICK STORMSHAK, Oregon State University, Corvallis
ANNE K. VIDAVER*, University of Nebraska, Lincoln

Staff

GREGORY H. SYMMES, *Study Director* (since August 1999)
MICHAEL J. PHILLIPS, *Study Director* (through July 1999)
LUCYNA KURTYKA, *Research Associate*
SHIRLEY B. THATCHER, *Senior Project Assistant*
NORMAN GROSSBLATT, *Editor*

* Resigned from committee

Foreword

In 1989 the Board on Agriculture of the National Research Council recommended that an expanded public investment be made through competitive research grants in agriculture, food, and the environment. The rationale for this recommended program, to be administered by the U.S. Department of Agriculture (USDA), was a perceived need "to revitalize and reinvigorate one of [America's] leading industries, the agricultural, food, and environmental system." The objective was to increase the generation of new knowledge in key issue areas, which could best be accomplished by selecting the highest quality research proposals through the use of peer review. In Fiscal Year 1991, Congress created the National Research Initiative (NRI), the expanded competitive grants program at USDA. Without accounting for inflation, this program is currently funded at a level slightly less than one fourth of that recommended by the 1989 Research Council report.

In 1995 Frank Press, my predecessor as President of the National Academy of Sciences, chaired a National Academies committee that examined the allocation of federal funds for science and technology. Among the central recommendations of this committee were that: (1) federal agencies should make allocation decisions based on clearly articulated criteria congruent with those of the President and the Congress, (2) the allocated funds should ensure that the U.S. achieves preeminence in select fields and is world class in all other major fields of science and technology, and (3) competitive merit review, especially involving external reviewers, should be the preferred way to make awards. The NRI is a program aimed at meeting these three important criteria.

The National Research Council is releasing this report, focused on improving and strengthening the NRI, following a study carried out by an expert and knowledgeable committee. Many of its members have been successful, as described in Appendix 4, in competing for peer review grants at NSF, NIH, and NRI. This committee has conducted a retrospective assessment of the quality and value of the NRI program, examined its science and technology priorities, and suggested changes for the future. Among the key findings and recommendations are: (1) a major emphasis of the program should continue to be the support of high risk research that has potential long-term payoffs, as well as benefits in training and education, (2) the proposals and awarded grants are generally of high quality, but scientists outside the traditional "food, fiber, and natural resources" disciplines need to be attracted to the program, (3) a more effective performance tracking system needs to be established to improve research accountability, (4) the priority setting process needs significant revision with creation of six standing scientific research review committees to identify critical issues and with special consideration given to important problems perceived by the public, and (5) a new NRI advisory board with representatives of NRI stakeholders should be established. The committee has also reemphasized the original NRI budgetary recommendation (adjusted for inflation) of the 1989 Board on Agriculture report.

As emphasized in the 1995 report of the Press committee, the federal government has played a pivotal role in developing the world's most successful system of research and development. Maintaining the vigor of this science and technology enterprise, of which the NRI is an important component, is essential to the nation's future. By making changes in the NRI program of the type recommended here, the US can attract many more outstanding young scientists to careers in these critical areas.

I thank the chair and members of the committee that produced this important report. We hope that it will help to make the NRI even more effective.

Bruce Alberts
Chair
National Research Council

Preface

Our nation faces daunting challenges to its food and fiber system and to the condition of our natural resources in the coming decades. Rapid increases in world population and the pressure on resources generated by increasing per capita consumption as a result of increasing per capita income challenge the very basis of our standard of living—our food, fiber, and natural-resource base. As the nation faces the challenges, new technologies and new information systems are changing the face of biologic research.

The US Department of Agriculture (USDA) has traditionally been the nation's primary public research engine in food, fiber, and natural resources. The National Research Initiative Competitive Grants Program (NRI) is a small part of that USDA research effort, but it accounts for a substantial portion of the nation's merit-based peer-reviewed fundamental research efforts in food, fiber, and natural resources.

USDA asked the National Research Council to review the NRI from four perspectives:

1) To perform a retrospective assessment of the quality and value of research funded by the program.

2) To determine whether the science and technology priorities in the major NRI programs are defined appropriately.

3) To assess how NRI activities complement other USDA programs, programs of other federal agencies, and state programs in the private sector.

4) To recommend the nature and content of changes for the future.

To respond to the request, the Research Council established the Committee on Evaluating the USDA National Research Initiative (NRI) Competitive Grants Program. The charge is broad. To assess the quality and value of research, the committee gathered data from the literature and solicited informed opinions from a broad spectrum of researchers and administrators who have long experience with the NRI and other federal institutions involved in merit-based peer-reviewed research. The committee did not have the time or expertise to review each funded activity and assess its specific quality and value. Assessing quality and value of fundamental research is difficult in any program without many years of hindsight. Had we discovered important concerns about the quality or value of research funded by the NRI during our deliberations, we would have had to revisit our approach to the charge. We did not have to do that.

It is a daunting prospect to undertake the evaluation of priorities for any research endeavor that covers as wide an array of topics as that in the NRI portfolio. We addressed the question of whether grant-setting priorities met congressional mandates. We studied the priority-setting process itself in much detail and have made recommendations for change. We chose not to define priorities per se although we have suggested a wide range of research concerns that, in our judgment, the NRI needs to address for the nation's food, fiber, and natural-resources agenda. Our recommendations on structure and process should, however, allow the NRI to put a detailed priority-setting process into place, and that should result in a comprehensive agenda for NRI research.

Addressing the question of complementarity required defining how the NRI food, fiber, and natural-resources activities fit into the nation's other public and private research activities. We endeavored to do that.

Finally, we chose to give item 4—recommending the nature and content of changes for the future—the broadest of interpretations. Information garnered during this study required that we address the funding and structure of the NRI to respond fully to the charge. A Research Council committee recommended dramatic increases in funding of the NRI in the 1989 report *Investing in Agriculture*, and the interim Research Council review in 1994 reiterated that position. We revisited the subject and responded with recommendations on both structure and funding. Our recommendations reaffirm and extend the earlier Research Council vision for fundamental merit-based peer-reviewed research in food, fiber, and natural resources.

Substantial recommendations are made to strengthen the NRI itself and, by strengthening the NRI, to enhance the nation's peer-reviewed research efforts in food, fiber, and natural resources, which, if successful, will prepare us for the coming decades.

We are, indeed, unprepared for many of tomorrow's food, fiber, and natural-resources challenges. There is much to be done to avert catastrophe if the projected increases in world population are realized. I am convinced that the adoption of the committee's recommendations by Congress and the executive branch will dramatically improve the nation's preparedness to address the challenges.

Thomas N. Urban
Chair
Committee on An Evaluation of the National
Research Initiative Competitive Grants Program

Acknowledgments

The committee is extremely grateful to numerous people who gave of their time and expertise to provide data and other input during the study process and the development of this report. It is difficult to provide an exhaustive list of those who contributed to this effort, but the committee wishes to thank the following, who provided input during public meetings: Kenneth Barton, Monsanto Life Sciences, St. Louis, Missouri; Robert Bremel, Gala Design, Sauk City, Wisconsin; Anthony Cavalieri, Pioneer Hi-Bred International, Inc., Johnston, Iowa; Harold Coble, Council for Agricultural Science and Technology, Raleigh, North Carolina; Jack Eberspacher, National Association of Wheat Growers, Washington, DC; Noah Engelberg, Office of Management and Budget, Washington, DC; David Ervin, Wallace Institute, Greenbelt, Maryland; Kellye Eversole, National Corn Growers Association, Chevy Chase, Maryland; Kirk Ferrell, Animal Agriculture Coalition, Washington, DC; Cliff Gabriel, Office of Science and Technology Policy, Washington, DC; Karl Glasener, Tri-Societies, Washington, DC; Richard Herrett, Agricultural Research Institute, Washington, DC; Tracy Irwin Hewitt, C-FARE, Arlington, VA; Charles Jamison, National Corn Growers Association, Beltsville, Maryland; Andrew Jordan, National Cotton Council, Mamphis, Tennessee; Eileen Kennedy, US Department of Agriculture, Washington, DC; Victor Lechtenberg, Purdue University, West Lafayette, Indiana; David McKenzie,

State Agricultural Experiment Station, College Park, Maryland; Terri Nintemann, US Senate Committee on Agriculture, Nutrition, and Forestry, Washington, DC; Terry Nipp, AESOP Enterprises, Washington, DC; Kenneth Olson, American Farm Bureau Federation, Park Ridge, Illinois; Lyle Roberts, Illinois Soybean Association, Bloomington, Illinois; John Suttie, University of Wisconsin, Madison, Wisconsin; and Robert Zimbelman, Federation of Animal Science Societies, Bethesda, Maryland.

In addition, we would like to thank those who took time to meet with the committee throughout the study process: Colien Hefferan, US Department of Agriculture, Washington DC; Ted Hullar, Cornell University, Ithaca, New York; Ron Phillips, US Department of Agriculture, Little Rock, Arkansas; Michael Roberts, US Department of Agriculture, Washington, DC; and Sally Rockey, US Department of Agriculture, Washington, DC.

Two hundred eighty-eight persons took the time to fill out the survey that provided much useful information to the committee. We are grateful for the responses and for the many additional comments and ideas provided.

This report has been reviewed by people chosen for their diverse perspectives and technical expertise, in accordance with procedures approved by the National Research Council Report Review Committee. The purposes of this independent review are to provide candid and critical comments that will assist the authors and the Research Council in making the published report as sound as possible and to ensure that the report meets institutional standards of objectivity, evidence, and responsiveness to the study charge. The content of the review comments and draft manuscript remain confidential to protect the integrity of the deliberative process, but we wish to thank the following for their participation in the review: Roger Beachy, Danforth Plant Center, St. Louis, Missouri; John Burris, Marine Biological Laboratory, Woods Hole, Massassachusetts; Susan Cozzens, Georgia Institute of Technology, Atlanta, Georgia; Rodney Croteau, Washington State University, Pullman, Washington; Jackie DuPont, Florida State University, Tallahassee, Florida; Ronald Estabrook, University of Texas Southwestern Medical Center, Dallas, Texas; Jack Gorski, University of Wisconsin, Madison, Wisconsin; Edward Hackett, Arizona State University, Tempe, Arizona; George Hallberg, The Cadmus Group, Waltham, Massachusetts; Theodore Hullar, Cornell University, Ithaca, New York; John Shadduck, Shadduck Consulting LLC, Fort Collins, Colorado; Philip Smith, McGeary and Smith, Washington, DC; and Patrick Windham, R. Wayne Sayer and Associates, Washington, DC Although the reviewers listed above provided many constructive comments and suggestions, responsibility for the final content of this report rests solely with the committee and with Research Council.

The committee wishes to thank Study Director Michael J. Phillips (through July, 1999), Research Associate Lucyna Kurtyka, and Administrative Assistant Shirley Thatcher for their assistance during our deliberations and in preparing this report. Their organizational skills contributed enormously to the study process. We gratefully acknowledge the editorial work of Anne H. (Kate) Kelly on early versions of the report. We also thank Laura Boschini for her efforts in preparing the final report for publication.

Finally, the committee and the members of the Board on Agriculture and Natural Resources wish to extend special thanks to Gregory H. Symmes, study director since August 1999. It was our great fortune that Greg agreed to assist us in the completion of the report and that the Research Council's Commission on Geosciences, Environment, and Resources was willing to share Greg's expertise. Greg provided invaluable advice and direction during the final phase of the study. His experience, knowledge, energy, and persistence helped the committee through a long, and sometimes difficult, process. We deeply appreciate Greg's untiring and extraordinary efforts in seeing this study to a successful conclusion.

Table of Contents

Appendixes

Tables, Figures, and Boxes

TABLES

FIGURES

BOXES

Executive Summary

The nation's food, fiber, and natural-resources system has always evolved, but the pace of change is now more dramatic than ever. In the life sciences, new knowledge created by a better understanding of animal, human, microbial, and plant genomics is providing new opportunities to control pests and disease, enhance the quality and safety of food, improve nutrition, and increase productivity. Equally impressive advances are occurring in information technology, providing the opportunity to increase productivity, minimize environmental impacts, and fundamentally alter decision-making.

The ability of the United States to resolve challenges to the food, fiber, and natural-resources system by developing sustainable food and fiber production; enhancing food safety, quality, and nutrition; protecting an increasingly fragile environment; responding to predictable cycles of global warming; and developing alternative energy sources depends on the depth of public knowledge, the public availability of technologies, and the skill and insight to apply them.

The US Department of Agriculture (USDA) spends about $1.7 billion per year on research related to the nation's system of food, fiber, and natural resources, of which about $120 million is spent on merit-based peer-reviewed research funded by the National Research Initiative Competitive Grants Program

(NRI). The $1.6 billion that USDA spends on research through non-NRI programs is distributed noncompetitively through intramural research grants to USDA staff (which can include cooperative agreements with land grant universities and other organizations), formula funds to state agricultural experiment stations, and special grants for targeted initiatives and direct grants to states. This allocation system does not in itself necessarily reduce the quality or relevance of research, but it runs counter to practices at the National Institutes of Health (NIH) and the National Science Foundation (NSF) and to the general direction of most federal research practices for assessing research quality and relevance.

The NRI is the nation's primary merit-based peer-reviewed research response to challenges to its system of food, fiber, and natural resources. The potential for disease transfer between animals and humans; the use of crops as substitute sources of petroleum-based products; the advent of nutraceuticals (specific foods for the prevention or treatment of disease); preparation for and prevention of biologic terrorism; the environmental impacts of farming, food-processing, and forestry; and the improvement of the vitamin and mineral content of widely grown grains are just a few examples of important emerging research issues directly relevant to USDA's mission. Merit-based peer-reviewed research on such issues could have profoundly beneficial effects in the United States and the rest of the world, especially in developing countries.

HISTORY OF COMPETITIVE RESEARCH AT USDA

Competitive merit-based peer-reviewed grants at USDA were first authorized by Congress in 1977. Congress provided $15 million to start the program and mandated that it be open to any researcher who would submit a grant application. From 1977 to 1989, the program grew to $40 million per year. In 1989, the National Research Council called for expanding competitive grants in a new program with proposed annual funding of $550 million. Congress responded in the 1990 Food, Agriculture, Conservation, and Trade Act by authorizing annual spending of up to $500 million on a new competitive grants program within 5 years. Congress initiated the NRI in FY 1991 with an appropriation of $73 million. Annual funding for the NRI was increased to about $100 million in FY 1992 and remained at or near this level through FY 1998. In FY 1999, the NRI budget was increased to $120 million. Since its inception, the NRI has functioned as a pilot program to support high-quality research related to the nation's food, fiber, and natural-resources system.

ORGANIZATION

The NRI is in the Competitive Research Grants and Awards Management Division of USDA's Cooperative State Research, Education, and Extension Service (CSREES). It is governed by a Board of Directors that comprises the administrators of all the USDA intramural research agencies and the under secretary for research, education, and economics, who is the board chair.

The NRI has six divisions organized according to the six mandated programs authorized by Congress: Animal, Plants, Food and Nutrition, Marketing and Trade, Natural Resources and Environment, and Food Processing. The scientific staff consists of the chief scientist, division directors, program directors, and the rotating panel managers recruited from the research community to administer NRI review panels.

STUDY PROCESS

In 1997, USDA asked the National Research Council's Board on Agriculture (now the Board on Agriculture and Natural Resources) to conduct an independent assessment of the NRI program. Specifically, USDA asked the Board to (1) perform a retrospective assessment of the quality and value of research funded by the program, (2) determine if the science and technology priorities with the major NRI programs are defined appropriately, (3) assess how NRI activities complement other USDA programs and those of other federal agencies and state programs in the private sector, (4) recommend the nature and content of changes for the future. The Research Council appointed a 14-member committee in early 1998 to carry out this study.

To respond to USDA's four-point charge, the committee gathered impressions and systematic data on the performance of the NRI. The committee conducted a series of surveys and interviews and solicited testimony from several constituent groups. Former chief scientists, deans and directors of land grant and non-land grant universities, and recipients and nonrecipients of NRI grants were included in mail surveys as a first comprehensive effort to assess the functioning of the NRI. In addition, the committee devoted a full day to receiving testimony from interested stakeholder[1] groups. Every effort was made to gain the views of individuals or groups that had had contact with the NRI and were therefore knowledgeable as to its activities. The committee found a great deal of consistency in findings from the survey, interviews with the chief scientists, and testimony presented by stakeholders at a public workshop.

Early in the study the committee recognized that the NRI did not maintain a systematic record of direct research results (for example, publications, patents) or a running evaluation of the originality and significance of current applications and renewals. The committee therefore based its assessment of the "quality and value of research funded by the program" (its first task) largely on surveys, testimony, and its own experience. To supplement these subjective evaluations, the committee chose to expand the scope of its investigations to evaluate how well the NRI program has met the goals that were set forth in the 1989 NRC report and the original congressional authorization, some of which involve

---

[1]The term *stakeholder* is used here to refer to all individuals and organizations that have an interest in the operations and outcomes of the NRI.

organizational and funding issues. In adopting this expanded charge, the committee therefore has discussed a number of organizational and funding issues and has offered recommendations to help achieve the original goals for this program and to give it greater visibility within, and external to, USDA.

STATUS OF THE NATIONAL RESEARCH INITIATIVE

The committee found the NRI to have financed high-quality scientific work within congressional guidelines. The committee also found, however, that the program is in danger of languishing. Program size, grant duration, grant size, and a low overhead allowance have led to reduced application numbers. Applicants are primarily from traditional food, fiber, and natural-resource sciences. A key goal of the program—to attract scientists from outside the traditional food complex—has not been achieved.

Furthermore, the committee found that traditional stakeholders in the NRI are losing confidence in the health and direction of the program. Uneven and opaque internal procedures, funding allocation processes, and priority-setting patterns have reduced the desirability of the program in the eyes of potential applicants in and outside the traditional food-research complex.

Finally, the location of the NRI within the USDA organizational structure suggests that the USDA and Congress place a higher priority on formula funds, special grants, and intramural research than on extramural, merit-based peer-reviewed research. Expectations of increased funding for the NRI generated by two National Research Council reports (in 1989 and 1994) and the 1990 congressional authorization have not been met. That has generated frustration in the food, fiber, and natural-resource research community and has had an adverse effect on the acceptance of the NRI as a strong research program.

The committee reiterates the extraordinary importance of public merit-based peer-reviewed research in food, fiber, and natural resources. In the committee's opinion, past public research and current private activities cannot meet the needs that are being created by population growth, climate change, and natural-resource deterioration or the challenges related to food safety and nutrition and to the growing convergence of foods and medical research.

THE NRI'S MISSION

A successful grants program contains elements of value, relevance, quality, fairness, and flexibility. The committee found that the proposals to the NRI and the research conducted by scientists who receive NRI grants are of high quality. That finding is based on the results of the committee's survey of applicants, awardees, administrators of land grant institutions, and industry; the views of former chief scientists and individuals from federal agencies; and the personal perspectives of committee members and their colleagues. Through conscientious stewardship, the NRI has been successful in generating fundamental and applied research and fostering the development of future scientists with strong backgrounds in food, fiber, and natural resources.

The committee recommends that a major emphasis of the NRI continue to be the support of high-risk research with potential long-term payoffs. Much of this research would be classified as fundamental in the traditional use of this term. The NRI also should continue to emphasize the importance of multidisciplinary research.

The NRI program is credited with important contributions to fundamental and applied research. The distinction between fundamental (or basic) and applied research often is unclear in the food, fiber, and natural-resources sector, however. Instead of classifying research arbitrarily as fundamental or applied, it should be thought of as on a continuum with short-, medium-, and long-term objectives identified in any research area. The committee believes that a major emphasis of the NRI should continue to be the support of high-risk research with potential long-term payoffs—the type of research that is unlikely to be funded through other research programs in USDA, other federal agencies, or the private sector. The committee also encourages the NRI to continue to emphasize multidisciplinary research because the problems in the food, fiber, and natural-resources system demand multidisciplinary approaches and collaboration.

The committee recommends that the NRI continue to emphasize its mission of training and education.

The training and education of graduate students and postdoctoral researchers attributable to the NRI program have been valuable. Although grants have been small and of short duration, training appears to have been a major use of NRI funds among university researchers. "Strengthening grants"[2] provided by the NRI program have had a major impact on the careers and productivity of faculty who otherwise would not receive federal grant support. Furthermore, NRI staff have been successful, particularly in view of the organization's limited resources, in organizing several vehicles to promote public understanding of research in food, fiber, and natural resources.

RESEARCH ACCOUNTABILITY

The committee recommends continuing the process of merit-based peer review as the most effective method of competitively distributing funds for research in food, fiber, and natural resources.

The committee views the NRI as a model of merit-based peer-reviewed research in USDA. Because it uses a competitive review process to rank

[2]Strengthening grants are made available to faculty of small- and medium-sized academic institutions or institutions in USDA-EPSCoR (Experimental Program for Stimulating Competitive Research) entities who have not received NRI awards during the previous 5 years.

proposals, however, the NRI remains outside the mainstream USDA culture of formula funding. The successful operation of the peer-review system in the NRI accounts for the high quality of the projects funded. Stakeholders in the food, fiber, and natural-resources system hold the NRI's peer-review process in high esteem. Some survey respondents indicated that the NRI merit-based peer-review process was as fair as and perhaps more responsive than the review process of other federal research agencies.

The committee recommends that a more effective performance-tracking system be established to improve research accountability.

The committee believes that the NRI could improve its record by documenting the value of research that it funds. The NRI does not keep a definitive record of patents and publications resulting from NRI research. Nor is there a running evaluation of originality and significance of current applications and renewals. Although the committee has found based on its surveys that funded applications are of high quality, the NRI lacks a tracking system of critical factors needed for self-evaluation or effective reporting of research accomplishments to outside groups, which would create a feedback system to establish value.

Every federal research agency faces important challenges in measuring outcomes of research projects, and the NRI is no exception. The committee concluded that the quality of research supported by the NRI is high, but it was unable to scrutinize individual projects extensively because of the absence of a tracking system tailored to tying projects to outcomes. A standardized tracking system needs to be implemented for the NRI program. Such a system would be beneficial both for tracking outcomes and for making the NRI's programs more transparent to stakeholders. The National Research Council has recently released a report, *Evaluating Federal Research Programs,* on accounting for federal outcomes as part of the Government Performance and Results Act mandate. The NRI should use the recommendations in that report.

The committee recommends implementation of an internal information system that generates data on current operations of the NRI.

The committee found it difficult to follow year-to-year changes in funding areas and to generate numbers to measure effort by project and category outcome. The committee's requests for information generated more work by the NRI professional staff than should have been required. The committee believes that those problems were due to deficiencies in the underlying information system itself.

The committee recommends that the NRI Web site be more readily accessible to allow the location of research projects and results with the use of issue-oriented key words and technical terms that are accessible and understandable to all stakeholders.

A number of recommendations reflect directly on the NRI's ability to reach both traditional and new stakeholders. But the needs for transparency, access to the current research agenda, and documentation of past outcomes suggest a substantial expansion in communication strategy. A Web site could be linked to nontechnical summaries, technical abstracts, impact statements, and publications and to a catalogue of current and past funded projects. Such data and communication could be maintained for 10 years to build a timely, comprehensive, and searchable record of research impacts generated by NRI funding.

PRIORITY SETTING AND ORGANIZATION

The committee has concluded that the priority-setting process of the NRI needs substantial revision. The committee found that parts of the process used by the NRI staff seem unstructured, appear to be unevenly administered across NRI divisions, and are not explicitly linked to the goals and other strategic planning elements of the Research, Education and Economics Mission Area. Changes in program areas and priorities appear to have occurred primarily in response to the urging of vocal stakeholders rather than as the result of a deliberative priority-setting process. Mechanisms are not well established to evaluate the effectiveness of NRI-funded research as time passes and progress occurs or to delineate how key research outcomes correlate with guiding research goals. The priorities of the NRI do not appear to be linked closely with the priorities of the Agricultural Research Service (ARS) and the Economic Research Service (ERS), perhaps because the potential cross-functional nature of present research programs is not fully appreciated in either the ARS or NRI administration.

The committee believes that an improved priority-setting process should involve independent input from scientists and informed members of the public. The priority-setting process also should allocate more of the NRI's funds by issue, not by research category. The committee believes that changes in the NRI's organization need to be made. Most important, USDA needs to find a way to enhance the position of extramural research in USDA and to encourage NRI priority-setting to reflect national priorities more clearly.

The committee offers the following recommendations to improve the priority-setting in and the overall effectiveness of the NRI. Other solutions are possible; ultimately it will be up to USDA, and possibly Congress, to decide how best to address these problems.

The committee recommends that six standing scientific-research review committees be assembled to identify critical issues in each research area. The committee further recommends that the current 26 programs be eliminated and replaced with an issue-based agenda across the purviews of the six committees.

Some NRI divisions have been relatively stable programmatically since their inception, whereas others have seen many program starts and stops. The subdivision of the NRI's six main research areas into 26 programs solely by research "category", in the absence of an overall strategic plan, might have been partly responsible for a lack of critical mass among the NRI's natural stakeholders, particularly because the recommended increases in research funding to $500 million did not materialize.

Several short-term changes in program direction (over 4- to 6-year time frames) have occurred in research areas that would otherwise need about 8–10 years to have an impact. The stop-start nature of some NRI funding commitments over its short history indicates that the NRI has been unable to sustain funding support for some high-risk areas with long-term payoffs—the types of research for which the NRI is ideally suited. The lack of a clear perception of the logic of annual requests for proposals across all 26 programs could be partly responsible for the NRI's inability to attract increased research budgets for its programs. A more logical, priority-setting process that relates the NRI's research programs to USDA goals and emerging issues in the food, fiber, and natural-resources system might be effective in demonstrating more clearly the importance of NRI-supported research and lead to increased research budgets.

The committee recommends that the research review committees give special consideration to important problems perceived by the public at large—such as alternative energy, healthfulness of food, food safety, and nutrition (issues at the consumer end of the food system)—in addition to the more traditional emphases on productivity, rural economies, and environmental protection.

The likely outcome would be a better distribution of research funds across the entire food, fiber, and natural-resources system and a research agenda more closely aligned with public concerns. The NRI research agenda would thus become more forward-looking and issue-driven.

The committee recommends that a cooperative formal goal and strategy process be instituted in the context of the NRI's role in federal food, fiber, and natural-resources research programs.

The NRI generally complements other USDA activities and does not duplicate other federal research efforts. The NRI actively participates in cross-agency funding opportunities to ensure complementarity of research efforts, but it clearly follows rather than leads in such efforts. Apart from memoranda of understanding and interagency coordination provided by the National Science and Technology Council, no process exists for establishing formal relationships with other federal agencies or for consulting and using stakeholder groups.

NIH, NSF, the Department of Energy (DOE) and NRI form the backbone of the nation's merit-based peer-reviewed research effort in food, fiber, and natural resources. The NRI is the nation's only merit-based peer-reviewed research program that focuses explicitly on challenges to its system of food, fiber, and natural resources. A comprehensive strategy that required coordination among congressional committees—particularly those with jurisdiction over USDA, NSF, and NIH programs and budgets—would allow an expanded NRI food, fiber, and natural-resources agenda to be coordinated with complementary work funded by NIH and NSF.

The committee recommends that the NRI and other competitive USDA research programs be moved to a new Extramural Competitive Research Service (ECRS) that would report to the under secretary for research, education, and economics.

The location of the NRI as one component of the Competitive Research Grants and Awards Management Division, rather than on an organizational level equivalent to USDA's two main research agencies (ARS and ERS) suggests that USDA and Congress place a higher priority on formula funds, special grants, and intramural research than on extramural, merit-based peer-reviewed competitive research. The committee believes strongly that unless extramural competitive research is given the same stature organizationally as formula-funded and intramural research in USDA, it will remain difficult for the NRI program to achieve its mission.

The committee believes that the NRI has suffered as a program in an agency—CSREES—that is also responsible for defending and allocating formula funds and special grants. Intramural research is represented by ARS and ERS, which report directly to the under secretary for research, education and economics, as does CSREES. The committee strongly recommends that extramural competitive research be given an organizational stature that would allow it to compete effectively for resources with formula funds and special grants and to participate directly in USDA's high-level priority-setting process.

The committee recommends the establishment of a new Extramural Advisory Board (12–14 members) that represents NRI stakeholders and has a non-USDA chair.

Funding has been unevenly allocated among the NRI's divisions since its initiation. No substantial changes in the proportions of funding allocated to the divisions have occurred, even though the nature of food, fiber, and natural resources has changed since 1991. Funding allocations do not appear to have distinguished between traditional and emerging areas in food, fiber, and natural resources.

The current NRI Board of Directors provides necessary administrative oversight of the NRI program and can be used to link the NRI with USDA's other research organizations. The Board of Directors is not responsible for providing guidance on scientific or technologic priorities, providing a forum for stakeholder concerns, or measuring research outcomes and evaluating NRI operations. An external advisory board of some type is necessary to handle those responsibilities.

The Advisory Board would advise and assist the chief scientist in identifying fundamental issues and future strategies to meet the greatest needs. It would represent scientists and engineers, deans of land grant and non-land grant institutions, industry across the entire food and fiber system, commodity and farm groups, consumer groups, and 1890 colleges. Ex officio members would include select program managers at NIH and NSF and the NRI chief scientist. Board members would serve 3-year terms on a staggered, rotating basis with a maximum of two terms. The board would be appointed by the secretary of agriculture.

In the committee's opinion, an external Advisory Board is critical to the successful functioning of the NRI. Stakeholder contact, the advocacy of extramural research inside and outside USDA, measurement of research outcomes, and continuing evaluation of NRI operations (including the peer-reviewed project-selection system) would ensure thoroughness, objectivity, and transparency. A visible, mandated external Advisory Board would bring renewed energy and focus to an expanded NRI effort and would provide Congress with an objective appraisal of NRI efforts.

The committee recommends that the position of chief scientist be a full-time, permanent 5-year position, with an option of one 5-year renewal, chosen by the secretary of agriculture with the consultation, recommendation, and advice of the newly created NRI Advisory Board. The chief scientist would be the administrator of ECRS.

The current responsibilities of the NRI chief scientist are equivalent to a full-time position. A part-time revolving chief scientist cannot meet the strategic-planning, priority-setting, and communication needs of an effective NRI. Although past chief scientists have done excellent work, having a part-time chief scientist impedes continuity in accountability and leadership and counters successful long-range planning and followup and consistent stakeholder involvement.

The necessary duties of the chief scientist-administrator of ECRS, in addition to those now assigned within the NRI, would include directing the program and developing a definitive strategic plan and advocacy for the NRI program. The chief scientist could also take the lead in changing the culture of the NRI from a program-based to an issue-based research agenda. The full-time chief scientist would report directly to the under secretary and would play a major role in setting the nation's federal food, fiber, and natural-resources research agenda.

The committee recommends that each of the six mandated areas of research emphasis be led by a half-time associate chief scientist with a 2-year rotation. Each associate would be a scientist from a visible and productive outside research program.

In recent years, the NRI staff has been stretched to cover its responsibilities, and this has increased the burdens of communication and timeliness on NRI staff at all levels and on the scientists who serve as ad hoc reviewers and panel members. The proposed rotation system would allow the chief scientist to recruit a flow of intellectual capital and would provide a mechanism for obtaining input from the population of researchers served by the NRI. The full-time chief scientist plus the six associate chief scientists would have the time and resources to carry out long-term analyses of research needs in the context of issues rather than programs, as is now the case. This recommendation highlights the importance of establishing and maintaining a scientifically based research agenda. The associate chief scientists would complement the division directors, program managers, and volunteer panel leaders.

A number of factors could account for the fact that USDA's research agenda has struggled over the last decade. The committee understands current budget constraints and understands that the implementation of some of its recommendations would increase personnel and operating costs. We believe strongly, however, that substantial changes are needed to ensure the future success of merit-based peer-reviewed research in food, fiber, and natural resources.

FUNDING

The committee recommends that grant awards be immediately increased to an average of $100,000 per year (total costs) over 3 years.

NRI research grants are much smaller and shorter than grants supporting similar types of research at NSF, NIH, and DOE. Continued underfunding of NRI research grants relative to those of other federal research agencies will tend to discourage new researchers outside the traditional food and fiber system from applying for NRI grants—one original goal of the NRI. It might also cause highly qualified scientists who have received NRI support to apply for research funds from other sources and even redirect their research away from issues important to the food and fiber system. That could lead to a decrease in the overall quality of food, fiber, and natural-resources research.

The proposed increase would solidify the stakeholder foundation of the NRI and prepare it to receive additional funds. The committee recognizes that without an increase in the NRI's total budget (as recommended strongly by this committee), the increase in size and duration of grants would reduce the number of grants and perhaps cause hardship among investigators who have depended on NRI funding to sustain their research programs. However, continued underfunding of individual research grants will reduce the aggregate impact of the NRI's competitive funding. The number of current proposals is lower than in the past, and stakeholder support appears to be waning. An increase in the size and duration of grants would enable the scientific community to attack issues in food, fiber, and natural resources by preparing proposals that require multi-investigator and multidisciplinary teams of researchers. Increased size and duration of grants would allow researchers to carry out projects as planned without narrowing their scope to fit a shorter period and smaller amount. Finally, increased size and duration of grants would attract new, creative proposals from researchers who are now outside the traditional food and fiber system. The latter was one of the key reasons for instituting the NRI, and it continues to be a worthwhile objective. To achieve it, the NRI must provide realistic funding levels to continue to attract the best and the brightest students and investigators to food, fiber, and natural-resources research.

The NRI should benchmark the funding level and duration of its grants to those of the other federal merit-based peer-review agencies that support research. NSF and NIH support competitive research projects in some of the same basic science and engineering areas as the NRI, that complement food, fiber, and natural-resources research. The challenge is to keep the best intellectual capital engaged in the NRI's scope of issues.

The committee recommends that the NRI's overhead limit be immediately replaced with indirect-cost standards that are used by other federal research agencies.

When it established the NRI program in 1991, Congress imposed a 14% limit on the amount of indirect costs that can be charged as a percentage of the total award.[3] The 14% limit was replaced by a 19% limit[4] in FY 2000 as part of the Agricultural Research, Extension, and Education Reform Act of 1998. Although the increase from 14% to 19% reduces the gap between overhead rates on NRI grants and rates on grants awarded by other federal agencies, overhead rates for most academic and private-sector research institutions are significantly higher than the 19% limit currently allowed. Average overhead rates for NSF's Biology Directorate, for example, are approximately 45% of the modified total direct costs of the award—nearly double the NRI limit. The committee is not aware of any other federal merit-based peer-reviewed research program with such a congressionally mandated limit on overhead rates.

Presumably, the motivation for setting such a limit was to increase the percentage of NRI research funds spent on research activities. However, such a mandated cap on overhead may have a negative effect on the NRI program because it causes some institutions (especially those from outside the traditional applicant community) to discourage their researchers from submitting proposals to the program. Because the committee did not address this issue in its survey, it was not able to estimate the magnitude of this effect on the NRI program.

[3]This limitation is equivalent to 0.16279 of the total direct costs of an award.
[4]This limitation is equivalent to 0.23456 of the total direct costs of an award.

However, the committee is aware of one research institution that prohibits its scientists from submitting proposals to the NRI because the low overhead rates do not cover the true institutional costs associated with such research and because its auditors require consistency among all incoming grants. Other institutions discourage their researchers from submitting proposals by requiring that the researchers (or their departments) use other funds to make up the difference between mandated low overhead rates and the established rates used by other federal agencies. This is especially problematic for smaller institutions where researchers do not have the flexibility to balance low-overhead grants against other sources of unrestricted funds. These factors also may have a disproportionate impact on institutions (or departments) from outside the traditional food, fiber, and natural-resources system because they do not have a historic association with USDA and may be less willing to accept a low overhead rate that is unique to USDA-sponsored research.

The committee believes that Congress could help broaden the scope of NRI researchers beyond the traditional food, fiber, and natural-resources system—one of the original goals of the program—by allowing the NRI to use the same negotiated overhead rates used by other federal agencies. This action, together with the increased grant amounts recommended previously, would make the NRI a more attractive source of funding to all institutions and researchers and could encourage proposals from researchers from outside the traditional food, fiber, and natural-resources system

The committee recommends that by 2005 the NRI budget be increased to a level equivalent (adjusted for inflation) to the $550 million recommended by the NRC in 1989—but only if recommended changes in priority setting, documentation, and organization are put into place.

Inadequate funding of the NRI has significantly limited its potential and placed the program at risk. A substantial increase in funding will ensure a robust and high quality public research effort that can significantly transform the nation's food, fiber, and natural resources system in response to critical needs in agricultural productivity, environmental health, and societal well-being.

In its 1989 report *Investing in Agricultural Research,* the NRC called for expanding competitive research within the USDA and establishing the NRI, with a proposed funding increase to $550 million within one year, if possible. Congress responded in 1990 by authorizing $500 million for the NRI by 1995, but the current program is only $120 million. The committee strongly re-affirms the previous NRC recommendation, and has estimated that the equivalent size of the NRI budget would be approximately $800 million in 2005. The committee believes that attaining this level would be an important step in re-energizing the national food, fiber, and natural resources research complex—which in turn, would result in major benefits to the nation. After reaching this budget level, the future growth of the NRI budget should be evaluated and compared with the growth in the budgets of complementary research programs in NSF, NIH, and

DOE, as suggested in the committee's earlier recommendation to benchmark the amount and length of NRI grants against such research programs.

The committee believes that the recommended increase in funding should take place incrementally as the various changes recommended earlier in this report are put into place. The ability to utilize large amounts of new funding effectively will be compromised unless recommended changes to the priority-setting process and NRI's organization are implemented.

A NATIONAL FOOD, FIBER, AND NATURAL-RESOURCES RESEARCH COMPLEX

If implemented, the recommendations growing out of this third National Research Council review of the NRI (the other two were in 1989 and 1994) will re-energize the NRI and the nation's food, fiber, and natural-resources research complex and will give USDA the opportunity to rediscover its fundamental research roots—where it began 120 years ago. In the committee's opinion, the nation needs USDA to re-emerge as the research engine of the food, fiber, and natural-resources complex that has served the nation so successfully in the 20th century. There is no acceptable alternative. The food, fiber, and natural-resource system is too important and too fundamental to future national security and stability not to have its own research program that focuses explicitly on high-risk problems with potential long-term payoffs. The committee believes that an expanded and refocused NRI is the proper platform. Without a dramatically enhanced commitment to merit-based peer-reviewed food, fiber, and natural-resources research, the nation places itself at risk.

1

Introduction

Over the last 20 years, the US food, fiber, and natural-resources system has experienced dramatic changes. It has expanded rapidly to include health, safety, and environmental issues. Emerging technologies cast transgenic crops as pharmaceutical factories and soils as mitigators of atmospheric CO_2 increases. Improved understanding of animal, human, microbial, and plant genomics is providing new opportunities to control pests and disease, enhance the quality and safety of food, improve nutrition, and increase productivity. Equally impressive advances are occurring in information technology, providing the opportunity to increase productivity, minimize environmental impacts, and fundamentally alter decision-making. New discoveries and their applications are changing how business is done in the global food and fiber marketplace. Public-sector research has been at the heart of the nation's response to challenges to its food, fiber, and natural resources.

The predicted addition of 3 billion people to the world's population over the next 30 years—a 50% increase—could have adverse economic and social effects, especially in the food and housing sectors. The increasing population could have major effects on the world's limited supply of arable land and cause substantial environmental degradation, including the potential for global climate change. Spinoffs from US food, fiber, and natural-resources research to the developing world therefore could be even more important as population pressure intensifies.

Underlying the dramatic shifts in society's expectations of the nation's food, fiber, and natural-resources system has been a rapidly expanding research agenda based on discoveries in chemistry and biology. Some of the exciting advances that will drive future developments in this system include

- Discoveries in plant and animal molecular biology, in ecosystem science, and in plant and soil chemistry and biology.
- Development of information technology that allows food to be tracked from producer to consumer.
- More information about the connection between diet and the body's defenses against disease.
- Measurement of major economic relationships and their connection to institutional change and organizational structure in the food and fiber system.
- Genomic studies of agricultural crops, plant pests, and beneficial microbes.

The National Research Council's Committee on Evaluating the National Research Initiative believes that merit-based peer-reviewed research on such issues can have profoundly beneficial effects in the United States and the developing world. As the nation's primary merit-based peer-reviewed research response to challenges to its system of food, fiber, and natural resources, the US Department of Agriculture (USDA) National Research Initiative (NRI) competitive grants program should play an important role in such progress.

This report summarizes the current status of the NRI and offers a number of recommendations to improve its effectiveness. This introductory chapter provides a brief overview of the history of competitive research at USDA and the NRI itself, summarizes the results of prior reviews of the NRI, briefly describes the committee's study process, and provides a brief guide to the report.

BRIEF HISTORY OF COMPETITIVE RESEARCH AT USDA

The passage of the Hatch Act of 1887 established USDA as the first federal agency to sponsor extramural scientific research. A formula-based funding process based on each state's share of total rural and farm populations permitted the establishment of USDA Agricultural Research Service laboratories in several geographic locations and annual funding to state agricultural experiment stations. This approach to funding has provided considerable flexibility at the state level to use funds to address practical food and fiber problems and to build and maintain the local research infrastructure. Although formula funds have provided little support of fundamental research (research having no immediate application—see discussion in chapter 4), the combination of mission-oriented research, teaching, and extension in the land grant colleges provides a unique structure that rapidly transmits research results to the farm, student, and consumer level and has contributed to a dramatic expansion of food and fiber output in the United States.

In the last century, and particularly since World War II, a new approach to funding research developed in American universities outside the food, fiber, and natural-resources system. Funds were awarded for projects on a competitive, peer-reviewed basis rather than by geographic or economic formula. Many have cited the new approach as one of the principal reasons for the success of the research enterprise in the United States: the approach " has relied on an abiding faith in the superiority of a free market in ideas and entrepreneurial competition over top-down decision-making in ensuring the quality and efficiency of research efforts" (CED, 1998).

In 1969, the secretary of agriculture asked the National Research Council to sponsor a broad-based evaluation of the food and fiber research enterprise. The resulting report (NRC, 1972) acknowledged the historical strength of US food and fiber research but found that "far too much of the research is of low scientific quality" and that "agricultural research is suffering from an inadequate interaction with the basic disciplines that underlie it." Reasons for those deficiencies included the findings that "grossly inadequate support was given to the basic sciences that underpin agriculture" and that there was "inadequate opportunity for a free flow of ideas from the scientist to the funding source."

The National Research Council report (NRC, 1972) contained 20 recommendations to remedy those and other deficiencies in the food and fiber research enterprise, among them that USDA

> seek a greatly increased level of appropriations for a competitive grants program, which should include support of basic research in the sciences (biological, physical, social) that underpin the USDA mission,...available to scientists in the USDA, in land grant and non-land grant public universities or colleges, and in private universities or colleges, institutes, and other research agencies,...administered in such a way that research proposals are subjected to evaluation by peer panels,...[and funded to] approximate 20% of the USDA's research budget. [Pp. 49–50]

Five years later, the congressional Office of Technology Assessment (OTA) reiterated the need for a competitive grants program for food and fiber research in its report *Organizing and Financing Basic Research to Increase Food Production* (OTA, 1977), and Congress provided authorization for competitive research grants in the 1977 farm bill. That legislation established the USDA Competitive Research Grants Office (CRGO) and provided $15 million to start the program. From 1977 to 1989, the competitive grants program slowly expanded from $15 million to about $40 million per year.

In its 1989 report *Investing in Agricultural Research,* the National Research Council (1989) called for expanding CRGO into the NRI, with a proposed funding increase to $500 million. The Research Council report argued that a healthy NRI was necessary to address the three major issues facing US agriculture: its competitiveness, food safety, and environmental quality. The report justified its recommendation for increased funding on the basis that

> (1) The pervasive needs and problems require large amounts of new knowledge and technology for their resolution. (2) Agricultural research provides a high return on investment. (3) The agricultural research system, as presently funded, is unable to provide the necessary financial support for the quality, amount, and breadth of science and technology necessary to address the problems [P. 5].

Congress responded in the 1990 Food, Agriculture, Conservation, and Trade Act (FACTA) by expanding the competitive grants program into the new National Research Initiative, authorized to spend up to $500 million within 5 years (relevant sections of 1990 FACTA are reproduced in appendix A). The NRI was initiated in FY 1991 with an appropriation of $73 million. In later years, the appropriations fell far short of the authorized levels (see chapter 6).

FACTA called for four types of competitive grants:

- Single-investigator grants awarded to support a single scientist or coinvestigators working in the same discipline.
- Multidisciplinary team grants awarded to support collaborating scientists in two or more disciplines focusing on basic research.
- Multidisciplinary team grants awarded to support collaborating scientists conducting applied research, with technology transfer a major component of all such grant proposals.
- Institutional strengthening grants awarded to support an institution for the improvement of its research, development, technology-transfer, and education capacity through the acquisition of special research equipment and the improvement of agricultural education and teaching.

In authorizing appropriation of funds for the NRI, Congress stipulated distribution to categories of grants as follows: in each fiscal year, 30% for fundamental and applied multidisciplinary work; at least 20% for mission-linked systems research; and at least 10% for strengthening grants and awards to faculty at small and middle-sized institutions that have not been successful in their quest to obtain competitive grants. Mission-linked research is more applied and provides scientific understanding needed to solve current, identified problems of importance to food, fiber, and the environment. Mission-linked research can provide information and technology that is transferable to users and which can be related to a product, process or practice. With respect to strengthening grants, no more than 2% of the appropriated funds can be used for equipment grants. The overhead rate for the grants would be 14 percent.[1] No more than 4% of the fiscal year appropriation can be used by the secretary of agriculture for costs of administering the NRI.

---

[1]Congress increased the overhead rate from 14% to 19% in FY 2000 in the Agricultural, Research, Extension, and Education Reform Act of 1998.

NRI ORGANIZATION

The NRI is in the Competitive Research Grants and Awards Management Division of the USDA Cooperative State Research, Education, and Extension Service (CSREES). The NRI is governed by its Board of Directors, which consists of the administrators of all the USDA intramural research agencies and the under secretary for research, education, and economics, who is the board chair.

The NRI has six divisions organized according to the six mandated programs authorized by Congress: Animal, Plants, Food and Nutrition, Marketing and Trade, Natural Resources and Environment, and Food Processing. Program directors are the responsible scientific staff, and rotating managers are recruited from the research community to administer NRI review panels. A more detailed description and analysis of NRI's organization are provided in chapter 6.

The NRI program description is drafted each year by the chief scientist and scientific staff; it is guided by the authorizing legislation and appropriation level and based on user-workshop reports, advisory committees, suggestions from panel members, and priority-setting documents, such as OTA and National Research Council reports (see chapter 4 for a more detailed discussion of priority-setting at the NRI). The resulting request for proposals is published in the *Federal Register* and distributed widely within the scientific community.

PRIOR REVIEWS OF THE NRI

The only prior fully external review of the NRI has been the 1994 Research Council report by the Board on Agriculture, *Investing in the National Research Initiative*. That report stated that "the board believes that it is yet too soon to conduct a comprehensive evaluation of the NRI, its program areas, and the benefits from the research it has supported. Although early results are indeed encouraging the NRI is only now on its fourth granting cycle." The report went on to indicate, however, that "today, the board finds that the NRI has yet to reach the potential envisioned for it" owing in large part to low funding, which had restricted the number and size of grants. As stated in the preface to the report, "ultimately, the board found the rationale for the establishment and vigorous expansion of the NRI more compelling than ever."

An overview of the NRI was published in *BioScience* in 1996 by A. Kelman and R. J. Cook (1996), former NRI chief scientists. They noted that six research subjects in which major scientific breakthroughs had occurred had been targeted initially for support by the CRGO, the NRI's predecessor, including plant-pest interactions, plant and animal genetic mechanisms, human nutrition, and animal diseases. The concentration of scientific advances now forthcoming in these and related subjects reinforces the importance of sustained support for research.

STUDY PROCESS

In 1997, USDA asked the National Research Council Board on Agriculture (now the Board on Agriculture and Natural Resources) to conduct an independent assessment of the NRI program. Specifically, USDA asked the Research Council to: perform a retrospective assessment of the quality and value of research funded by the program; determine whether the science and technology priorities in the major NRI programs are defined appropriately; assess how NRI activities complement other USDA programs, those of other federal agencies, and state programs in the private sector; and recommend the nature and content of changes for the future. The Research Council appointed a 14-member committee in early 1998 to carry out this study.

To respond to USDA's four-point charge, the committee gathered impressions and systematic data on the performance of the NRI. It conducted a series of surveys and interviews and solicited testimony from several constituent groups. Former chief scientists, deans and directors of land grant and other universities, and recipients of NRI grants, and others were included in mail surveys as a first comprehensive effort to assess the functioning of the NRI. In addition, the committee devoted a full day to receiving testimony from interested stakeholder[2] groups. Every effort was made to gain the views of individuals and groups that had had contact with the NRI and were therefore knowledgeable about its activities. The committee found a great deal of consistency in findings from the survey, interviews with the chief scientists, and testimony presented by stakeholders at the public workshop.

OVERVIEW OF REPORT

This report summarizes the results of the committee's analysis. Chapter 2 summarizes the value of food, fiber, and natural resources-research to the United States, focusing on economic contributions and rates of return of food and fiber research. The committee's analysis of the quality, value, fairness, relevance, and responsiveness of the NRI competitive grants program is presented in chapter 3. Chapter 4 presents the committee's analysis of the role and scope of the NRI, including its scientific objectives, its value in training and education, and its complementarity with other research activities. The committee's analysis of NRI's priority-setting process and its research priorities is given in chapter 5. The committee's analysis of organizational and funding issues is given in chapter 6. Chapter 7 presents the committee's recommendations to improve the effectiveness of the NRI program. Additional supporting materials are found in appendixes A through I.

---

[2]The term *stakeholder* is used in this report to refer to all individuals and organizations that have an interest in the operations and outcomes of the NRI.

2

Value of Food, Fiber, and Natural-Resources Research

Many studies have demonstrated the value of publicly supported research in science and technology. For example, the 1995 National Research Council report *Allocating Funds for Science and Technology* found that "the federal investments in [the US scientific and technical enterprise] have produced enormous benefits for the nation's economy, national defense, health, and social well-being" (NRC, 1995, p. 3). A report of the US Committee for Economic Development, *America's Basic Research: Prosperity through Discovery*, noted that "continued excellence in basic research is essential to America's prosperity and global leadership" (CED, 1998, p. 2). That committee observed further that the federal government had long been the most important source of funding of basic research; this is true especially for food and fiber research, which until World War II was the principal beneficiary of federal funding.

Congress also has reviewed national trends in research, as recently described in the 1998 House Committee on Science report, *Unlocking Our Future: Toward a New National Science Policy* (the "Ehlers report"). The report, written under the leadership of physicist and US Representative Vernon Ehlers, documents the importance and the "stunning payoffs" of the federal research investment in the US technology enterprise. The importance of applications of research findings in the physical and chemical sciences and engineering to telecommunication, defense, transportation, and health is duplicated in the importance of applications of research in biology, agriculture, and engineering to food, fiber, and natural resources—the focus of this report. Agricultural and physical sciences alike benefit immensely from understanding-driven or basic research, targeted basic research, and mission-directed or applied research (the terms used in the Ehlers report).

In general, 20th century research in food, fiber, and natural resources has contributed substantially—in both quantitative and qualitative terms—to the stability and prosperity of the US economy and to the broader world economy. In this chapter, the committee summarizes the economic value of this research to the US economy, discusses the impact of advances in life sciences, and provides an overview of trends in public and private-sector funding of food, fiber, and natural-resources research.

ECONOMIC CONTRIBUTIONS

The economic contributions of food, fiber, and natural-resources research are reflected in the contributions made by the food and fiber system to the growth of the US and global economies (Lipton et al., 1968; Daily et al., 1998). In 1996, US farming and its related industries accounted for $997.7 billion—13% of the gross domestic product—and employed almost 23 million people and 16.9% of the civilian workforce and 20% of the workforce residing in nonmetropolitan areas. Farming itself accounts for less than 1% of the gross domestic product and employs only about 1% of the US workforce, but it is just one link in the value-adding chain of input suppliers, capital providers, processors, transporters, service units, retailers, and others who produce and deliver food and fiber products to consumers.

The United States is the world's leading exporter of food and fiber products, with sales of $60.4 billion in 1996. Crops from more than 30% of US farmers' acreage are exported. USDA's Economic Research Service estimates that each dollar earned from exported food and fiber stimulates another $1.32 of output in the United States. US food and fiber exports alone were estimated to support about 859,000 full-time jobs in 1996.

Food and fiber exports also are important in narrowing the foreign-trade gap. In 1996, the food and fiber trade surplus was $26.8 billion; the non-food-and-fiber account was in deficit by $235.1 billion. The surplus adds to the strength of the US dollar, which helps to control inflation and moderates the prices of imported goods. The nature of food and fiber exports has changed substantially from mostly bulk commodities—such as grain, feed crops, and oil crops in the 1960s and 1970s—to mostly high-value items, such as meat products, fruits, vegetables, and beer and wine. In addition to exports, 1995 sales by foreign food-manufacturing affiliates of US companies totaled another $113 billion.

The high efficiency of production and delivery of US food and fiber enables Americans to spend only about 11% of their disposable income on food—the lowest rate of expenditure in the world. In contrast, estimated rates of disposable income spent on food are 17% in Europe, 30% in South America, and 51% in India (B. Meade, USDA, personal communication, March 23,

1999). The low cost of food frees consumer income for other uses, allows a cost-effective focus on food quality and safety and human nutrition, and cuts costs to US taxpayers for food stamps and related public-assistance programs.

The US food and fiber system has responded quickly and effectively to important long-term trends. Changing incomes, demographics, lifestyles, and consumer perceptions of relationships between health and diet are among those trends. The ethnic diversity of the US population has broadened the array of food products available to consumers. The need for convenience in food-purchasing choices has led to greater diversity of services in basic foods (processing and prepared food). Fast-food establishments, restaurants, and hotel dining have shifted the location and style of consumption. Concerns about safety and dietary issues have led to products that have improved health and safety attributes, including improved nutritional quality. Health and safety information is now transmitted in a more coordinated fashion through the stages of the food system because of increasing reliance on production contracts and vertical integration.

RATES OF RETURN FROM FOOD AND FIBER RESEARCH

Since the late 1950s, more than three dozen studies have estimated rates of return on public investment in food and fiber research in the United States (Fuglie et al., 1995; Alston and Pardey, 1996; Barry, 1997). The studies have, for the most part, found high real rates of return from most categories of applied and basic food and fiber research. The estimated returns on research typically range from 35% to 60% per year. Those rates are high relative to the government's cost of funds, relative to returns on alternative investments, and relative to private sector rates of return. Fuglie et al. (1995) summarized the aggregate returns to agricultural research and extension for the period 1964 through 1982 (see table 2-1).

Table 2-1 shows that the annual rate of return on research investment in agriculture was estimated to be about 41% between 1950 and 1982. Such a historically high rate of return before 1982 illustrates the powerful impact of well-managed, targeted food and fiber research. Since the first commercial introduction of a recombinant-DNA product (human insulin) in 1982, there has been a substantial change in the technology of agriculture and in chemistry and biology. Maintaining such an effect of research and return on investment in food, fiber, and natural resources will require focused and wise investments in the research enterprise that will catalyze advances in agricultural biotechnology and in fundamental biologic and engineering research applied to food, fiber, and natural resources.

The return on investment in food and fiber research includes not only returns to the technology developers that benefit from the research outcomes, but also the returns to farmers, agribusinesses, consumers, and other members of society that benefit from the research outcomes. Thus, food and fiber research return rates have a broader scope and are generally larger than private rates of return on shorter-term industrial projects, which tend to be 10% to 15%.

TABLE 2-1 Aggregate Returns on Public Investments in Agricultural Research and Extension

Study	Methodology	Period	Annual Rate Return, %
Griliches, 1964	Production function	1949–59	35–40
Latimer, 1964	Production function	1949–59	—[a]
Evenson, 1968	Production function	1949–59	47
Cline, 1975	Production function	1939–48	41–50
Huffman, 1976	Production function	1964	110
Peterson and Fitzharris, 1977	Economic surplus	1937–42	50
		1947–52	51
		1957–62	49
		1967–72	34
Lu, Quance, and Liu, 1978	Production function, R&E[b]	1939–72	25
Knutson and Tweeten, 1979	Production function, R&E	1949–58	39–47
		1959–68	32–39
		1969–72	28–35
Lu, Cline, and Quance, 1979	Production function, R&E	1939–48	30.5
		1949–58	27.5
		1959–68	25.5
		1969–72	23.5
Davis, 1979	Production function	1949–59	66–100
		1964–74	37
Evenson, 1979	Production function	1868–1926	65
White and Havlicek, 1979	Production function	1929–72	20
White, Havlicek, and Otto, 1979	Production function	1929–41	54.7
		1942–57	48.3
		1958–77	41.7
Davis and Peterson, 1981	Production function	1949–74	37–100
White and Havlicek, 1982	Production function, R&E	1943–77	7–36
Lyu, White, and Lu, 1984	Production function	1949–81	66
Braha and Tweeten, 1986	Production function	1959–82	47
Huffman and Evenson, 1989	Production function	1950–82	41
Yee, 1992	Production function	1931–85	49–58

[a] Not significant.

[b] R&E gives estimated rate of return on combined research and extension expenditures. Otherwise, estimate is for research alone.

Source: Adapted from Fuglie et al. (1995).

The research returns reflect several other key characteristics. First, the research benefits generally occur over long periods (for example, up to 40 years). Second, specific research outcomes are relatively risky, especially when high payoffs are concentrated in a few major breakthroughs. Third, the research returns are magnified by stimulating technology adoption and further research in other countries, economic sectors, and industries.

The rates represent the returns on primarily production-based research involving plants and animals. The returns typically do not include the costs of externalities attributed to research, such as possible environmental degradation, adjustment costs of displaced labor, and other adverse effects on human health, communities, and families. Also not included in these returns are the significant contributions of economics and other social science research, which provide additional value. Examples of social-science research outcomes are economic and social policy analyses, decision support and forecasting information, institutional innovations, and new organizational structures in food and fiber production and distribution.

IMPACTS OF ADVANCES IN LIFE SCIENCES

Largely within the last decade, food and fiber research investments by the private sector have increased from a historical level of 2% to 4% of gross sales to 10% or more—a level that is more typical of value-added products than of traditional agricultural commodities. The trend reflects the reality of today's high research costs, largely the result of expensive technology not available 20 years ago. The development and application of biotechnology clearly illustrate this expense. Not only is the technology expensive, but its development often requires a multidisciplinary approach. For environmental technologies and others with no immediate proprietary application but widespread public payoffs in the long term, funding falls exclusively to the public sector. Moreover, the public sector is increasingly responsible for training the students that are needed by industry to use new technology. Assigning relative contributions of public and private funds in support of research is difficult.

The development of new technologies applicable to food and fiber has led to new relationships between the research and regulatory arms of the US Department of Agriculture and between USDA and other regulatory agencies. For example, genetically engineered plant and animal products now fall under the jurisdiction of the Food and Drug Administration if the engineering leads to substantially altered products. Products that contain new proteins, fats, or carbohydrates or that have greater potential for allergenicity than existing varieties must pass rigorous premarket review and must be appropriately labeled when brought to market.

Similarly, new technologies that affect how food and fiber production influences air, soil, and water quality are leading to new relationships between food and fiber research and the regulatory agencies responsible for environmental protection—in particular, the Environmental Protection Agency and its state-level equivalents. In some cases, such as high-density animal production, new technologies might increase the risk of environmental degradation; in other cases, such as precision agriculture, new technologies promise opportunities for improved environmental stewardship. All advances in technology place additional demands on the research enterprise apart from the discovery and development of the advances themselves.

PUBLIC-SECTOR AND PRIVATE-SECTOR RESEARCH FUNDING

The food, fiber, and natural-resources sciences held a privileged position until World War II. As late as 1940, almost 40% of federal expenditures for research and development ($29 million of $74.1 million) was allocated to USDA intramural and state experiment-station research (Mowery and Rosenberg, 1989). World War II transformed the federal research system. First, the government contracted large amounts of research to the private sector. That shifted much federally financed research, particularly defense-related research, to industry. Since World War II, about 75% of all federal R&D expenditures have gone to the private sector (Mowrey and Rosenberg, 1989). Second, the war spawned huge increases in federal R&D spending. National-security concerns were often the principal drivers. Social issues and priorities also motivated the expansion of federal R&D investment, including the Great Society programs, environmental concerns, public health, and recently concerns about the international competitiveness of US industries. Until the late 1970s, the United States spent more on research than all other industrialized countries combined (Mowery and Rosenberg, 1989).

After World War II, other federal agencies received a greater proportion of federal research funding relative to USDA. Because defense-related research dominated federal research spending, the Department of Defense, Department of Energy, and National Aeronautics and Space Administration have accounted for a large share of federal research obligations (about 70% in 1998). However, university-based research also received a large boost from the creation of the National Science Foundation (NSF) in 1950 and the expansion of the National Institutes of Health (NIH). NSF and NIH greatly expanded federal support for university research and for the universities' research infrastructure. In 1998, NSF and NIH together accounted for almost 22% of all federal research obligations and over two-thirds of the federal research obligations for universities and colleges (NSF, 1999). By 1998, USDA expenditures for research were about 2% of all federal research spending, and about 2% of federal support for university research was for food, fiber, and natural-resources research (NSF, 1999).

The government's role in supporting food and fiber research has had to adapt to the rising involvement of the private sector in research and development. The post-World War II period has witnessed a large increase in the private sector's contribution to food and fiber research. Several factors have spurred private industry's interest in food and fiber research, including scientific advances in molecular biology, increased market opportunities, and stronger intellectual property rights to biologic inventions. Between 1960 and 1994, private sector food and fiber research expenditures more than tripled in real terms. Today, the private-sector invests more in food and fiber research than do the federal and state governments combined (figure 2-1).

Those research expenditures mask a major shift in the type of research conducted in the private sector (figure 2-2). In 1960, the responsibilities of public and private research were clearly drawn. More than 80% of private research was for improving farm machinery or developing new food products or processing methods, and public research concentrated on increasing yields of crops and livestock. Since then, the private sector has developed a large research capacity in subjects long dominated by the public sector, such as plant breeding. By 1996, nearly 21% of private research was devoted to increasing crop and livestock yields by supplying farmers with improved crop varieties, animal breeds, feeds, and pharmaceuticals. Those trends suggest continuing challenges of overlap between the public and private sectors in some kinds of food and fiber research.

The dramatic growth in private investment in food and fiber research might have overshadowed the nation's historical public research agenda. The recent explosion of private investment in the food and fiber system is, however, built on the preceding long-term public effort. There is no reason to doubt that the importance of public-sector research to industry is any less for food and fiber than for other kinds of research. A 1997 patent-citation study—which found that 70% of patent-application citations were of public-sector research (Narin et al., 1997)—illustrates well the symbiotic linkage between public and private research investments. The public sector provides innovative and creative research that could take considerable time for commercial development or might not be undertaken at all.

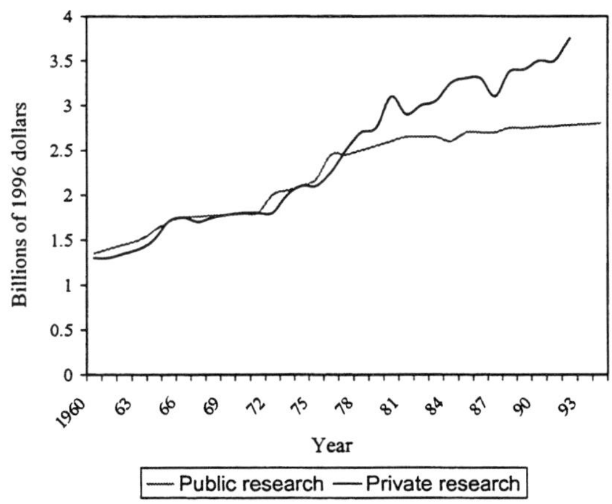

FIGURE 2-1 Food, fiber, and natural-resources research expenditures in the United States, 1960–1996

Sources: Based on Klotz, Fuglie, and Pray (1995), US Department of Agriculture, Economic Research Service, Private-Sector Agricultural Research Expenditures in the United States; public research data derived from US Department of Agriculture Inventory of Agricultural Research

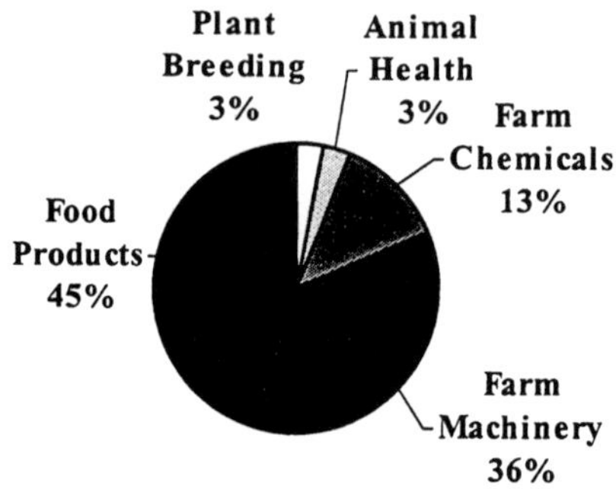

1960: 206 million

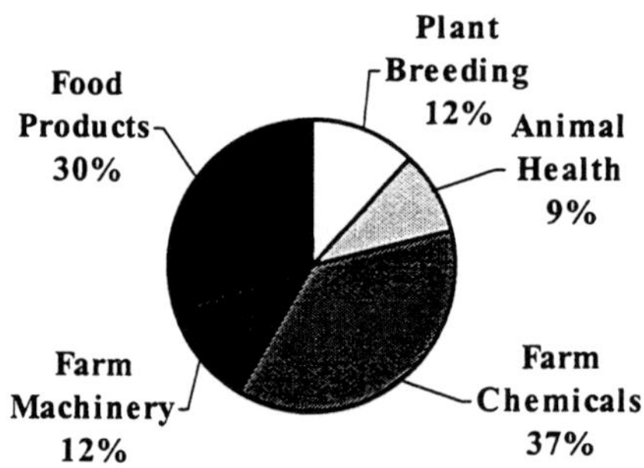

1996: 3.9 billion

FIGURE 2-2 Research expenditures (in nominal dollars) by food and fiber industries, 1960 and 1996.

Source: US Department of Agriculture, Economic Research Service. Data based on Klotz, Fuglie, and Pray, 1995

3

The National Research Initiative Competitive Grants Program

The defining attributes of a successful competitive grants program are quality, fairness, relevance, and flexibility (Chubin and Hackett, 1990; Chubin, 1994; Kostoff, 1997a,b). *High-quality* research is novel, valuable, feasible, technically sound, and on occasion elegant. In practice, the quality of basic research is conveyed by the publication of research results in a peer-reviewed scientific journal. The quality of applied research might be demonstrated by a patent or by the successful implementation of research results, perhaps as new management practices, new products, new institutional arrangements, or new public policies. High-quality research also impacts science and technology themselves—that is, on the direction and development of the scientific enterprise and its technical implementation—an attribute closely related to relevance.

Fairness refers to the likelihood that a proposal will be evaluated with strict adherence to a set of evaluation criteria related to the quality and relevance of the proposed research. Race, sex, age, geography, and institutional affiliation must be effectively ignored when one is evaluating a proposal. Fairness also means that the review process must be open to independent examination and that each proposal is considered seriously and appropriately by a well-qualified group of reviewers. In practice, this works by ensuring that reviewers and panel members are broadly representative of the entire scientific community in a given field.[1] A fair process also ensures that grant applications are solicited from as wide a variety of applicants as possible.

The remaining two attributes of a successful competitive grants program are relevance and flexibility. A *relevant* grants program provides funding for research that will most effectively further the goals of the program and meet national needs. *Flexibility* refers to the program's capacity to shift in response to emerging fields of research. Almost by definition, emerging fields are highly relevant. However, flexibility also should be intrinsic to the research enterprise as a whole. Achieving flexibility can be difficult because of institutional inertia—the addition of individual programs adversely affect the resources remaining for other programs. Thus, a mechanism for periodically evaluating and revising programmatic areas is crucial in a successful competitive grants program.

Other attributes of a successful program are related to specific practical aspects of the program's implementation. For example, the program must give awards of sufficient size, duration, and number to attract high-quality scientists and support important research. If the awards are too small or too short, many highly qualified scientists are likely to ignore the program in favor of other funding sources. Similarly, grants must be numerous enough to attract high-quality scientists, especially those at the beginning of their research careers. Grant acceptance rates below 10% suggest low chances for success and discourage many scientists from participating as either grant writers or reviewers. At very low funding rates, the effort expended by scientists in writing unsuccessful applications exceeds that of the scientists whose research is supported. Some have argued that such a program is a net burden rather than an asset to the scientific community as a whole (Chubin, 1998). Clearly, there are tradeoffs in the management of any research program (Chubin, 1994; Baldwin and McCardle, 1996). Implementation issues are analyzed in more detail in chapter 6.

QUALITY AND VALUE

Quality and *value* are terms commonly used to rank types of activities, and research is no exception. Specific metrics can be used to assess quality; alternatively, testimonials can be obtained from various sources to tap perceptions of quality. The latter approach generally was used by this committee to assess the quality and value of NRI-supported research. The former approach is addressed later in a committee finding on evaluation of quality and program accountability.

Evaluation of research has been a long-term challenge for the scientific community (NRC, 1998). In assessing the value of fundamental research, the private sector largely avoids such standard tools as return on investment and

---

[1]Government science agencies use peer-review in many ways. For additional information on the use of peer-review, see *Peer-review Practicies at Federal Science Agencies Vary*, General Accounting Office, 1999.

instead focuses on scientific merit and accomplishments of the researcher in fields relevant to the funding unit. The value of research in academe also includes the learning experience for the investigator and the student seeking answers to compelling questions. Value can also reflect the intended use of the research. Government agencies motivated by the Government Performance and Results Act use various approaches to measure value (Kostoff, 1997b). The Department of Energy Office of Basic Energy Sciences uses scientific excellence, relevance to energy future, stewardship, and program management. The National Science Foundation (NSF) uses expected outcomes, including discoveries at the frontiers of science, connections between discoveries and their use, and development of a diverse globally oriented workforce of scientists and engineers. The National Institutes of Health (NIH) considers both broad outcomes—such as understanding of biologic functions and behaviors, and improvement in prevention, diagnosis, and treatment of diseases—and specific descriptions of the known and unknown to understand and improve human health as evidence of research value.

Measuring the Quality and Value of NRI-Supported Research

The measurements of quality and value of NRI-supported research that the committee reviewed or generated included surveys of applicants, awardees, and institutions (see appendixes B and C); testimony from chief scientists, representatives of the private sector, government agency staff, and other constituents; interviews with NRI staff (see appendix D) and NSF officials; and the experience of members of the committee (appendix E) and their colleagues. The numerical quality indicators included the proportion of applications funded and successful renewal rates (appendix F). Documentation of successfully completed projects and their use and application added to the overall assessment (see boxes 3-1, 3-2, 3-3 and appendix G).

With regard to proposal quality, most of the surveys and testimonials agree in spirit with Harold D. Coble, representing the Council for Agricultural Science and Technology, who testified before the committee that panel reviewers consistently rated proposal quality as excellent. Similarly, past chief scientists unanimously concluded that NRI proposals were of high quality.

Funded NRI projects generally have come from the top of the priority pool established by the review panels. For example (this is typical of most categories), the category "Plant Responses to the Environment" received 1,196 applications over the 8-year period 1991–1998. Of those, 52 (4%) were considered outstanding and were funded; 208 (17%) were regarded as having "high priority", and 178 (86%) of them were funded; and 254 "medium-priority" applications (21%) were received, and 59 (23%) of them were funded. Only four "low-priority" projects out of about 300 so classified were funded over the 8-year period. An additional 341 applications were judged as "having some merit" or "do not fund" and were not funded. Over the 8-year period, 24% of applications were funded at some level. A total of $259 million in applications was received in the "Plant Responses to the Environment" category over the period. The funds requested for approved grants totaled $59 million; but only $40 million was actually awarded. In all, only 15% of the $259 million requested was awarded (NRI Program Office, personal communications, September 1, 1998).

In the committee's experience, high-quality applications generally lead to high-quality research. John W. Suttie, past president of the Federation of American Societies for Experimental Biology, testified that the quality of research selected through peer-review has always been high. Richard A. Herrett, of the Agricultural Research Institute, indicated that the NRI has been productive in providing new techniques. He cited the American Society of Plant Physiologists observation on the importance of NRI funding in developing environmentally benign insecticides. Robert G. Zimbelman, chairman of the Coalition of Funding Agricultural Research Missions, testified that "the record shows the NRI has supported very high quality research and the results have been meaningful." Kenneth E. Olsen, dairy and animal health specialist with the American Farm Bureau Federation, testified that "within the scientific community the NRI is well respected for top quality basic research." Tony Cavalieri, of Pioneer Hi-Bred International, Inc. testified that

> we believe [NRI] is a very sound program and, in fact, may be the most effective example of USDA using their research money effectively. They have been effective in funding important work and in funding researchers who can do the work....As far as the quality of funded research...it is obviously among the best work done in plant sciences.

In individual discussions with the committee, several NSF personnel cited NRI research as excellent. A previous National Research Council report on the NRI (NRC, 1994) attested to the quality of the research by noting the "consensus among NRI staff and panel members and managers that 'good to high' characterizes the overall quality and relevance of the proposals being received and that the quality has been increasing each year" (p. 21). The report also stated that "the contribution of the NRI extensive review process to the quality of science should not be overlooked" (p. 20). In short, the stewardship of the NRI has been unquestionably high.

BOX 3-1
Ethanol from Biomass:
An Example of a Significant Scientific Advance from NRI Research

If renewable biomass sources are to supply tomorrow's energy needs, cost-effective technologies are needed. Researchers at the University of Florida have been making significant strides toward removing barriers to ethanol production from biomass. Using biologic approaches, Lonnie Ingram and colleagues are laying the foundation for ethanol production from biomass that is less costly and less capital-intensive. The general approach has been to develop different microorganisms in which useful traits for cellulose hydrolysis and sugar metabolism are combined with genes for ethanol production. With NRI support, genetically engineered *Escherichia coli* were developed to produce ethanol from all the monomer sugars that can be derived from plant cell walls.

The resulting strain, KO11, has been used to ferment hemicellulose sugars derived from several sources of biomass. Later awards have led to the integration of the ethanol production genes from *Zymomonas mobilis* into the chromosome of *E. coli* and the engineering of *Klebsiella oxytoca* for the simultaneous saccharification and fermentation of cellulose.

More recent efforts are seeking to improve these microorganisms further, for example, by engineering the secretion of an *Erwinia* endoglucanase in *E. coli* and *Klebsiella oxytoca*. The aim is to reduce the requirement for supplemental cellulases from fungi, which are costly. Other research is to develop ethanol-producing biocatalysts with increased resistance to toxic products generated during the chemical hydrolysis of lignocellulose components. These compounds, sugar and lignin degradation products, currently must be removed by an expensive multistep process.

Work continues to make the biomass conversion to ethanol more competitive, but a milestone has been reached. On October 20, 1998, BC International broke ground in Jennings, Louisiana, for a commercial scale plant to produce ethanol from agricultural waste. The plant, which has the capacity to produce 20 million gallons of ethanol per year, will run on bagasse (a residue from sugarcane refining) but has flexibility to use other feedstocks as well. This first-of-its-kind plant is based on the genetically engineered KO11 bacterium developed by Dr. Ingram and colleagues.

Competitive Research Grants Office and NRI Awards to Dr. Ingram

1986 Constructing of Lactose Utilizing Strains of *Z. mobilis,* $102,000, 3 years.
1988 Genetic Engineering of Alcohol Production in *E. Coli,* $110,000, 3 years.
1990 Genetic Engineering of Bacteria for Ethanol Production, $120,000, 3 years.
1992 *In Vitro* Analysis of Plant- Pathogenic Mycoplasma Like Organisms, $180,000, 3 years.
1995 Genetic Engineering of Bacteria for Ethanol Production, $180,000, 3 years.
1998 Engineering Bacteria for Fuel Ethanol Production, $179,000, 3 years.
1998 Advanced Ethanologenic Biocatalysts for Ligno Cellulose Fermentations, $298,935, 3 years.

Patents

US Patent 5,000,000, Ethanol production by *Escherichia coli* strains co-expressing *Zymomonas* PDC and ADH genes, Mar. 19, 1991.
US Patent 5,028,539, Ethanol production using engineered mutant *Escherichia coli*, July 2, 1991.
US Patent 5,162,516, Cloning and sequencing of the alcohol dehydrogenase gene from *Zymomonas mobilis*, Nov. 10, 1992.
US Patent 5,424,202, Ethanol production by recombinant hosts, Jun. 13, 1995.
US Patent 5,482,846, Ethanol production in Gram-positive microbes, Jan. 9, 1996.
US Patent 5,602,030, Recombinant glucose uptake system, Feb.11, 1997.
US Patent 5,821,093, Recombinant cells that highly express chromosomally-integrated heterologous genes, Oct. 13, 1998.

Source: US Department of Agriculture, National Research Initiative Competitive Grants Program Office, Personal Communication, March 1999.

BOX 3-2
Porcine Reproductive and Respiratory Syndrome:
An Example of a Significant Scientific Advancement from NRI Research

Porcine reproductive and respiratory syndrome (PRRS), formerly known as mystery swine disease, first became apparent in the United States in 1986. In the next few years, it affected a majority of herds in every state that raised pigs. Economic losses were estimated at $250–300 per breeding-age female, so a typical 600-sow farm could lose $150,000–180,000 per outbreak, excluding other costs. By 1990, PRRS had appeared in Europe and was well on the way to becoming a global epidemic. The National Pork Producers Council considered the disease "the most important animal health problem affecting pigs".

Scientists knew that PRRS was caused by a highly contagious virus but understood little about how it was transmitted and where it replicated inside the animal. With funding from the NRI and the National Pork Producers Council, David Benfield, at South Dakota State University, and James Collins, at the University of Minnesota, began to address these issues. They identified the primary targets of the virus: the lung, heart, blood vessels, and lymph nodes. They also discovered that the virus is transmitted from pig to pig by close contact, such as nose-to-nose touching, by exposure to bodily secretions, by semen to female pigs, and from mother pig to fetus. An additional finding--that PRRS virus replicates in an unknown primary target tissue and is then released into the bloodstream--was especially important because it suggested that a vaccine could be successful in fighting the disease. In partnership with private industry, Benfield and Collins developed a vaccine using a weakened form of the virus. They also developed monoclonal antibodies for use in laboratory tests of pig serum or tissue samples. These antibodies allow quick, accurate, and economical diagnosis of the disease, thus reducing treatment costs and producer losses.

Although the vaccine and diagnostic resources described above are widely used today, PRRS continues to challenge producers and scientists as new strains of the virus emerge. The research funded by the NRI represents significant progress in understanding and combating the disease and laid the groundwork for continued advances in controlling this important animal-health problem.

NRI Awards to Drs. Benfield and Collins

1992 Pathogenic Mechanism of Swine Infertility and Respiratory Syndrome Virus, $150,000.

1995 Mechanisms of Persistence of Porcine Reproductive and Respiratory Syndrome Virus, $210,553.

1998 Rushmore Conference: Mechanisms in the Pathogenesis of Enteric Diseases, $5,000.

Patents

US Patent 5,677,429, Monoclonal antibodies to the mystery swine disease virus, Oct. 14, 1997.

US Patent 5,683,865, Vaccine for mystery swine disease and method for diagnosis thereof, Nov. 4, 1997.

US Patent 5,846,805, Culture of swine infertility and respiratory syndrome virus in simian cells, Dec. 8, 1998.

Source: US Department of Agriculture, National Research Initiative Competitive Grants Program Office, Personal Communication, March 1999.

BOX 3-3
Reduction in Fertilizer Use Profits Farmers and the Environment: An Example of NRI Research with Great Promise

Pamela Matson and colleagues, at Stanford University, used NRI funds from the Forest/Range/Crop/Aquatic Ecosystems Program to counteract the consequences of the Green Revolution. Although this greening of the world has indeed increased crop yields (through irrigation and application of nitrogen fertilizers), it has done so in conjunction with unwanted greenhouse gas accumulations (specifically nitrous oxide), increased tropospheric levels of ozone and acid rain due to increases in nitric oxide, and deposition of nitrates from soils into freshwater and marine ecosystems, often resulting in eutrophication. However, Matson and colleagues' research confirms that high yields are possible if less fertilizer is used, resulting in both lower application costs for farmers and lower social costs for the environment.

The research took place in the Yaqui Valley of Sonora, Mexico, a major wheat-producing region that has helped to foster the Green Revolution with high productivity, using fertilizers and irrigation. The experiment included a control where no fertilizer was added, a conventional farming treatment currently in use in the region (adding nitrogen at 250 kg/ha), and three alternative farming methods that used less fertilizer before irrigation.

All but the control had a yield of about 6 tons/ha, or 2.4 tons/acre, but there were significant differences in the amount of nitrogen released into the soil and air. The best alternative method, which applied 28% less nitrogen than the conventional method, resulted in a 69% reduction in total nitrogen loss and an approximate savings of $55–75/ha or $22–30/acre to the farmer. Thus, this alternative method is not only environmentally friendly, but also agronomically feasible and economically more desirable.

NRI Award (directly related to this research)
1994 Forest/Rangeland/Crop/Aquatic Ecosystems Program (now Ecosystem Science Program), $431,112 for 3 years.

Publications (directly related to this research)
Matson P., Naylor R., Ortiz-Monasterio I., 1998. Integration of environmental, agronomic, and economic aspects of fertilizer management. *Science* 280:112–115.

Source: US Department of Agriculture, National Research Initiative Competitive Grants Program Office, Personal Communication, March 1999.

Novelty and Significance of Research

The committee had difficulty in evaluating the novelty and impact of the 8-year research portfolio of the NRI. This difficulty is not peculiar to research organizations. Discussions with NSF indicated similar challenges. Clearly, however, a number of novel and significant results have occurred (see appendix G). The committee concluded that, although evidence suggests that NRI-funded research is novel and significant, detailed documentation is lacking.

A definitive record of patents and publications resulting from NRI research is not available, nor is a continuing evaluation of current applications and renewals as to their originality and impact. Information from the private sector about applications of the results of NRI-funded research also is lacking.

FAIRNESS

Fairness is the keystone of any successful competitive grants process. Participants must know that their proposal will be treated in a procedurally impeccable way that gives no advantage to a competitor and that is open to scrutiny without betraying confidential information. This delicate balance is maintained principally by staff, although panel managers can be instrumental. The committee believes that high-quality merit-based peer review is an essential component of a fair competitive grants program because it subjects all proposals to systematic scrutiny by knowledgeable specialists and requires ratings of quality and feasibility that constitute valuable advice to agency staff. In the following sections, the committee describes the NRI's peer-review process and then evaluates its effectiveness.

The NRI Peer-Review Process

The NRI peer-review process is administered by "panel managers" who work as short-term consultants with US Department of Agriculture (USDA) staff and bring extramural scientific credibility to the program and its operation. Panel managers are instrumental in recruiting reviewers for the panels and reinforcing the perception that cutting-edge research is being considered for NRI funding. The participation of some 15,000 proposal reviewers and panelists (many with experience on both NSF and NIH panels) in the NRI peer-review process over 8 years attests to the community's commitment to maintaining program quality (NRI Program Office, 1998). The rules and guidelines for panel composition are given in box 3-4.

After examining all proposals submitted to the specific programs, the panel managers and the program directors determine the scientific expertise needed to evaluate each proposal and then assign proposals to appropriate panelists and ad hoc reviewers. Ad hoc reviewers are selected to extend the panelists' scientific expertise. Careful attention is paid to avoid conflict of interest during reviewer selection. Each reviewer, whether a panel member or an ad hoc reviewer, is advised to treat the proposal with confidentiality. Before the panel meeting, copies of each proposal are mailed to the panel members and to the ad hoc reviewers for evaluation. Panelists write their own reviews of the proposals; then copies of the ad hoc reviewers' comments and evaluations are mailed to panelists. This procedure allows the panelists to develop their own views regarding the proposals before reading the ad hoc reviews but gives them time to study the ad hoc reviews before the panel meets.

Each proposal is evaluated by a primary and a secondary reviewer and by a third panel member who serves as "reader". The primary and secondary reviewers provide written reviews. Evaluation criteria include scientific merit of the proposal, qualifications of the investigators, and relevance of the proposal to program goals and to long-range improvements in the sustainability of US food and fiber. The same evaluation criteria are applied to renewal applications, with particular consideration of progress during the previous award period.

BOX 3-4
Rules and Guidelines for Panel Composition

In general, each panel is formulated as a true peer panel, representative of people eligible to apply to the particular program themselves. To the extent possible, the panel is representative of

- Demonstrated expertise in a relevant discipline.
- All major regions of the United States.
- Many types of institutions—land grant, private, public, industrial, and governmental institutions.
- Levels of academic position (academic and scientific maturity).
- Balance between new panelists and those who have previously served.
- Diversity in sex and minority status.

Panel continuity is achieved by requesting part of a panel to serve in the following year. Continuity of the review process is also ensured by asking some of the ad hoc (mail) reviewers for a repeat review if a proposal is revised and resubmitted in the following year.

A tentative list of panel members is approved by the chief scientist before any contact is made with prospective panel members. After the panel is assembled, it is submitted through the division director and chief scientist to the deputy administrator of competitive research grants and awards management of Cooperative State Research, Education and Extension Service for administrative approval.

At the panel meeting, the primary reviewer of each proposal presents a synopsis of his or her review and evaluation to the panel. The secondary reviewer also presents his or her evaluation of the proposal. The third reviewer (the reader) is provided the opportunity to make additional evaluative comments but does not furnish a written review. The primary reviewer also summarizes the ad hoc reviewers' evaluations. The proposal is then discussed by the panel and ranked by consensus. Ranking involves placing a proposal in one of six funding priority groups (outstanding, high, medium, and low priorities for funding; some merit; and not to be funded) and then ranking its merit relative to that of other proposals in the priority group.

Before concluding, and after all proposals are reviewed, the panel reconvenes for an intense reassessment of the proposal rankings. The reranking session allows the panel to check and affirm that each proposal is properly and fairly ranked with respect to relative scientific merit. The degree of reranking varies, depending on the particular panel. The consistency of this process across panels is ensured by oversight and communication among the program directors and by visits of the program directors to panels in other programs.

The program director and panel manager record the final ranking of the panel. Proposals are generally funded according to this ranking until program funds are depleted. The best proposals receive most of the funding, but program managers on occasion are able to fund lower-ranked proposals with "strengthening" funds that can be used to award standard strengthening, postdoctoral, and new investigator grants in rank order.[2] All award decisions are reviewed by the division director and the chief scientist. All applicants receive an anonymous critique of their proposals consisting of all reviewer comments.

Because panels for each program meet annually, nonrecipients of awards have little opportunity to appeal in the year of submission. However, they may resubmit their proposals in the next year and include a rebuttal to the reviewers' comments.

As a consensus body exempt from the Federal Advisory Committee Act, review panels can speak with a single voice. This attribute lends credence to a panel's recommendations and to the funding decisions made by NRI staff. By retaining 30%-40% of members from the previous year, the review panels embody the collective wisdom of research communities. Yet they must be sensitive to accusations of functioning as "old-boy" networks. They vary, too, in their willingness to advise staff on such issues as where to cut the budget in high-ranked proposals. Some panels refuse to discuss budgets; others routinely weigh such considerations.

Evaluation of the NRI Peer-Review Process

By and large, the committee's survey reveals that recipients and nonrecipients of NRI awards consider the NRI review process fair and effective (see appendix C). Most of those surveyed viewed the NRI as using a fair peer-review process to select proposals for funding, including 97% of awardees, 74% of nonrecipients, and 84% of participants in land grant institutions (appendix C). Respondents with review-panel experience in other competitive grants programs were especially complimentary of the NRI peer-review process.

The four former chief scientists interviewed for this study also expressed great confidence in the peer-review and evaluation process (appendix C). Similarly, despite some criticisms of the length of the review process, panel members representing NRI stakeholders generally viewed the process as satisfying and fair (appendix C). Overall, scientists, panel members, and administrators judged the fairness of the peer-review process as exceptional, and they considered peer-review the best way to assess projects and distribute funds.

---

[2]See chapter 4 for a discussion of NRI's "strengthening" awards.

RELEVANCE AND RESPONSIVENESS

A relevant grants program provides funding for research that will further the goals of the program and meet national needs. The committee's survey indicates that nearly all respondents (awardees, nonrecipients, and those in land grant institutions and industry) think that the program has contributed to the NRI mission of generating fundamental and applied research and fostering the development of future scientists with strong backgrounds in food and fiber (appendix C). It should be noted, however, that substantial fractions of respondents indicated that the NRI contribution has been less than they expected (16% of recipients; 27% of nonrecipients; and 43% of people in land grant institutions). Survey respondents, chief scientists, and those who testified before the committee repeatedly cited two main factors that have limited the NRI's ability to reach its full potential. First, nearly all expressed the view that the total budget for the program was inadequate and that awards were too short, too few, and too small (appendix C). Second, many of the respondents indicated that the NRI priority-setting process was not clear (appendix C). The committee discusses those two issues more fully in chapters 6 and 5, respectively.

In addition to being relevant, a successful research program must be responsive. How any program reflects intellectual developments while being sensitive to appropriations (an external constraint) and budgeting (an internal constraint) is a test of its responsiveness. Many communities see the NRI as *their* program; each has its own expectation of priorities. To be successful, NRI staff must manage relations with each community, and this is especially difficult in a constrained funding environment. One fundamental characteristic of a responsive research program is its ability to facilitate research in new and emerging fields through the creation of new programs and the consolidation of declining programs. NRI division directors and program managers who met with the committee assert that new panels and programs are created in response to the number of proposals received in a field, rather than in response to political pressure from commodity or other groups. The narrative history that NRI staff provided to the committee suggests, however, that the evolution of the six mandated divisions into the 26 current programs was overwhelmingly the result of upper-level management decisions (especially spinoffs of existing programs into the Agriculture Systems program) rather than the result of proposal submissions. The committee discusses this issue in more detail in chapter 5 as part of its analysis of the NRI priority-setting process.

A second important factor inherent in a program's responsiveness involves the allocation of funds to support different elements of the program's mission (such as basic research, mission-oriented research, and human development). For example, during their discussions with the committee, some NRI staff questioned whether the percentage of NRI funding spent on "strengthening" grants and the support of postdoctoral fellows and graduate students (currently 25%) has been adequate and whether such awards have been made appropriately. The committee discusses that issue in more detail in chapter 4 as part of its analysis of the role and scope of the NRI program.

A third important factor related to a program's responsiveness is how the program deals with the inherent tradeoff between the number of proposals funded and the average funding provided for each grant. NRI program staff have expressed a resistance to increasing grant size because such increases would require a decrease in the number of researchers receiving any support (assuming no increase in the program's budget). As a result, individual award amounts currently average 60% of requested amounts. NRI data show a decline in proposal submissions in recent years, which could be a result of growing doubts about the program's viability (see survey results in appendix C). The committee discusses that issue in more detail in chapter 6 as part of its analysis of NRI funding.

SUMMARY FINDINGS

A successful grants program contains elements of value, relevance, quality, fairness, and flexibility. The committee finds that the proposals to the NRI and the research conducted by scientists who received NRI grants are both of high quality. That finding is based on the results of the committee's survey of applicants, awardees, administrators of land grant institutions, and industry; the views of former chief scientists and individuals from federal agencies; and the personal perspectives of committee members and their colleagues. Documentation of successfully completed projects and their use and application was factored into the committee's assessment, as were the proportion of applications funded and successful renewal rates.

The committee believes that the NRI could improve its record by documenting the value of research funded. The NRI does not keep a definitive record of patents and publications that result from NRI research. Nor is there a running evaluation of originality and significance of current applications and renewals. Although the committee has found based on its surveys that funded applications are of high quality, the NRI lacks a tracking system of critical factors needed for self-evaluation or for effective reporting of research accomplishments to outside groups, which would create a feedback system to establish value.

The committee views the NRI as a model of merit-based peer-reviewed research in USDA. Because it uses a competitive review process to rank proposals, however, the NRI remains outside the mainstream USDA culture of formula funding.

Through conscientious stewardship, the NRI has been successful in generating fundamental and applied research and fostering the development of future scientists with strong backgrounds in food and fiber. The NRI program is, however, not as responsive or flexible as it could be. Proposal submissions have declined in recent years, owing in part to concern over the viability of the program and in part to the program's resistance to increase the size of grants because such an increase would come at the cost of supporting fewer researchers.

4

Role and Scope of the NRI Program

The role and scope of the NRI are broadly shaped by the legislative authority that established the program, which in turn was modeled closely after the 1989 National Research Council report *Investing in Research: A Proposal to Strengthen the Agricultural, Food, and Environmental System*. The NRI was authorized in section 1615 of the Food, Agriculture, Conservation, and Trade Act of 1990 (FACTA), which mandated a "national competitive research initiative" to be administered under the direction of the secretary of agriculture. Congress further stipulated that funds for the new program would be directed to "high priority research", defined as "basic and applied research that focuses on both national and regional research needs (and methods to transfer such research to on-farm or in-market practice)." The research was to be directed at six primary subjects: plant systems; animal systems; nutrition, food quality, and health; natural resources and the environment; engineering, products, and processes; and markets, trade, and policy.

The enabling legislation called for four types of grants: grants for principal investigators, grants for fundamental multidisciplinary teams, grants for mission-linked multidisciplinary teams, and grants for research-strengthening activities. *Investing in Research* identified grants in the first two categories as science-driven, that is, grants intended to advance science by supporting fundamental research relevant to food, fiber, and the environment. Grants in the second two categories were expected to be related to science and engineering questions of national importance linked to more-applied problems. All types were designed to strengthen US Department of Agriculture (USDA) research efforts.

In this chapter, the committee presents its views on the role and scope of NRI research within the bounds established by Congress —the NRI's niche within federal, state, and private-sector research in the food, fiber, and natural-resources system. We begin by discussing some of the scientific objectives of the NRI, such as supporting fundamental, applied, and multidisciplinary research. We then discuss some of the NRI's objectives related to training and education. Finally, we discuss the complementarity of the NRI to other federal research programs and the private sector.

SCIENTIFIC OBJECTIVES

Applied versus Fundamental Research

Research programs are commonly classified as supporting either fundamental (also referred to as basic) or applied (also referred to as mission-linked) research or both. The 1995 National Research Council report *Allocating Federal Funds for Science and Technology* defined basic research as having the following characteristics: "creates new knowledge; is generic, nonappropriable, and openly available; is often done with no specific application in mind; and requires a long-term commitment". The same report defined applied research as having the following characteristics: "uses research methods to address questions with a specific purpose; pays explicit attention to producing knowledge relevant to producing a technology or service; overlaps extensively with basic research; and can be short- or long-term".

Even though there is extensive overlap between the two types of research and the distinction between basic and applied research is often unclear (see also box 4-1), the original National Research Council proposal to establish the NRI and the enabling legislation for the program referred explicitly to the two general types. In particular, the enabling legislation required that "not less than 20% [of appropriations] shall be available to make grants for research to be conducted by persons conducting mission-linked systems research." More recently, the Agricultural Research, Extension, and Education Reform Act of 1998 increased the minimum for mission-linked research, requiring that "not less than 40 percent [of appropriations] shall be available to make grants for research … directly applicable to producers and agricultural production systems." NRI annual reports show that for the period FY 1991–1999 funding for mission-oriented research ranged from 26% to 50% of total research expenditures, and the data show a monotonic increase of about 4 percentage points per year in research perceived to be "mission-oriented" over this period.

NRI stakeholders seem to recognize the value of the NRI program in supporting both fundamental and applied research. The committee's survey shows that the NRI is widely perceived to be USDA's premier basic-research program, whereas USDA formula funds are perceived as supporting applied research (see appendix C). In the committee's queries of NRI awardees and unsuccessful applicants (appendix C), both groups overwhelmingly credit the program for making important contributions to fundamental and applied research, as do land grant and industry respondents. Comments from individual scientists reveal divergent opinions as to whether the program focuses too much on fundamental or applied research. These divergent views probably reflect a healthy mix of short-term and long-term application horizons.

BOX 4-1
Is It Basic or Is It Applied?

No a thousand times no; there does not exist a category of science to which one can give the name "applied science." There is science and there are the applications of science, bound together as the fruit to the tree which bears it.

Louis Pasteur, 1871

Although Pasteur was admirable in his emphasis on the unity of the basic and applied aspects of science, we continue to differentiate between basic and applied science in important ways. What is meant when we discuss basic and applied research, when the same study may be viewed as basic by some and applied by others? Basic and applied research can be distinguished by the intention of the work and the expected time frame for the use of the results. A study intended to solve an immediate practical problem is commonly viewed as applied; a study intended to uncover information is viewed as basic. Basic science may be considered to be curiosity-driven—knowledge is the intended product. Basic research is more long-term in perspective, and the utility of the results is likely to be more distant in time. What is viewed as applied research is intended to solve a practical problem, with a need to apply the information in the short term for a social or economic benefit.

Although the intentions of basic and applied research are quite different, the results may lead to important conclusions or utility that go beyond the original intentions. Many basic studies lead to unexpected discoveries that have immediate application, and applied studies often uncover new knowledge while in pursuit of a practical objective. Serendipity occurs in both kinds of research.

As recognized in the definitions of basic and applied research cited above, there often is no clear distinction between fundamental and applied science. That is especially true in food, fiber, and natural resources, where fundamental research complements applied research (for example, it is difficult to conceive of an advance in fundamental biology that might not have eventual application to the food system) and applied research often leads to further questions that fundamental research could help to answer. The committee therefore believes that it might be more useful to view NRI research along an expected-time-to-application gradient rather than to force each project to be classified as either applied or fundamental.[1] Recent developments in science and technology have

[1]The committee recognizes that such a change would probably require a change in the NRI authorization language, which prescribes a minimum of "mission-linked" research.

dramatically reduced the lag between laboratory research and useful products in many cases. It is now difficult to imagine a particular research project that might not contribute to some application within 10–20 years. Placing research along an expected-time-to-application gradient would distinguish research with an expected long-term payoff from research with a shorter-term horizon. It might also allow both to be distinguished from technology transfer (the movement of information from the public to the private sector) or product development. A technology-transfer or extension component might thus be encouraged for many projects other than those tagged "applied", and the extension community could be connected more directly to the research enterprise.

The committee does recognize that a potential negative effect of viewing NRI research along an expected-time-to-application gradient is the possibility that resources would be focused predominantly on near-term applications likely to produce short-term results rather than on longer-term research with a potential for higher payoffs. The observed increase in NRI research perceived as mission-linked over the last 8 years might reflect such a trend. The committee believes strongly that a major emphasis of the NRI should continue to be the support of long-term, high-risk research with potential long-term payoffs—the type of research that is unlikely to be funded through other research programs in USDA, other federal agencies, or the private sector. Much of this research would be classified as fundamental in the traditional use of the term.

Multidisciplinary Research

Multidisciplinary research has been defined as that "conducted by a team of collaborating scientists from two or more distinct science or engineering disciplines integrated into a single plan of study" (NRC, 1989, p. 13). Multidisciplinary research has been a long-time general goal in science to maximize the output and breadth of research applications. Accomplishing effective multidisciplinary research is not easy, because most research institutions have developed cultures that reward individual disciplinary accomplishments. In addition, multidisciplinary projects are often, by their nature, more expensive than research conducted by individual researchers. From 1993 on, by law, at least 30% of the value of NRI funded grants each year must be dedicated to multidisciplinary research.[2] The actual values ranged from 25%–34% from 1991 to 1994 (OTA 1995). In 1998, 43.4% of the NRI budget was defined as multidisciplinary.[3] The NRI appears to have delivered on its congressional mandate.

---

[2]Section 1615 of FACTA prescribed that "not less than 10% for fiscal year 1991, 20% for fiscal year 1992, and 30% for fiscal year 1993 and each fiscal year thereafter shall be available to make grants for research to be conducted by multidisciplinary teams."

[3]In the NRI's analysis, it was assumed that triagency (USDA, National Institutes of Health, and National Science Foundation) research was multidisciplinary.

TRAINING AND EDUCATION

Three components of the NRI program are directly related to training and education in the broad sense. In particular, NRI funds are used to develop future scientists through the support of graduate students and postdoctoral researchers, to strengthen small and medium academic institutions or institutions in USDA-EPSCoR (Experimental Program for Stimulating Competitive Research) entities, and to enhance public understanding of issues related to the nation's food, fiber, and natural-resources system.

Developing Future Scientists

The NRI supports about 425 graduate students[4] each year through its awards to project investigators. Responses to the surveys (see chapter 5) indicated that this aspect of the program is important and beneficial. In the words of one respondent, "many young scientists have started their career by NRI-supported research programs....NRI funds have played a major role in training, recruiting and retaining bright scientists in US agriculture." In fact, even with the small amount and brevity of grants, training might have become a major use of NRI funding among university researchers.

The NRI supports more than 300 postdoctoral researchers each year[5] including about 30 that receive direct postdoctoral fellowships.[6] Information on postdoctoral researchers supported by the NRI is presented in table 4-1. Both the total award amounts and the number of postdoctoral researchers increased from FY 1992 through FY 1996, followed by a general decrease from FY 1996 to FY 1999.

An important stated goal of the NRI is the training and support of young scientists—specifically, graduate students, postdoctoral fellows, and principal investigators who are in the initial or early stages of their scientific careers. How successful has the NRI been in accomplishing that goal, and how has the success been measured and documented? The one set of data available indicates that 5.7% of the total number of awards in 1997 went to new investigators. The range in different programs was 1.2%–7.3%. Each year, five to 14 grants, representing 0.16%–0.5% of the total number of awards, are given to postdoctoral fellows seeking their first grants (NRI Program Office, June, 1998).

No other data were available from the NRI on specific training of graduate or postdoctoral students, but traditional academic approaches generally use them extensively in research. That was confirmed in the survey of grantees, in which

---

[4]A graduate student is defined as a person pursuing an advanced degree, such as a Master's degree or doctorate.

[5]A postdoctoral researcher is defined as a person who has recently received a PhD in a field of science and is receiving further training in conducting research.

[6]Postdoctoral fellowships are awarded to highly promising researchers and can be used to obtain training in any research field. A fellowship is separate from research grant funds that can be used to support a postdoctoral researcher working on a specific research grant.

TABLE 4-1 NRI Support of Postdoctoral Researchers

Year	No. Grants	Funds Used for Supporting Postdoctoral Fellowships, $
1992	15	1,216,000
1993	20	1,635,000
1994	28	2,218,000
1995	31	2,451,000
1996	32	2,548,000
1997	27	2,362,726
1998	24	2,125,586
1999	20	1,744,503

Source: NRI annual reports for 1992, 1993, 1994, 1995, 1996, 1997, 1998, and 1999 (draft).

career development of graduate students was scored by 81 of 120 respondents as greatly affected and by 21 respondents as somewhat affected by NRI grants. A similar response (71 of 105 greatly and 19 of 105 somewhat) was reported with respect to career development of postdoctoral students. Even nonrecipients of awards believed that the NRI greatly or somewhat affected graduate and postdoctoral education. The survey also indicated an overwhelming belief that the NRI contributed to development of human resources in the food, fiber, and natural-resources research systems.

In general, the NRI has been successful in supporting the training of graduate students and postdoctoral scientists, given how small the program and its grants are relative to those of other agencies.

Strengthening Academic Institutions in the Food and Fiber System

Strengthening awards are made available to faculty of small and medium academic institutions or institutions in USDA-EPSCoR entities who have not received NRI awards during the previous 5 years. Small and medium institutions are defined as those with total enrollments of 15,000 or fewer that are not among the top 100 universities and colleges in receiving federal funds for science and engineering research.

USDA-EPSCoR entities—besides the District of Columbia, Puerto Rico, and other US commonwealths, territories, possessions and their successors—comprise states that have had a funding level from the NRI no higher than the 38th percentile based on a 3-year rolling average. In FY 1999, the NRI supported six career-enhancement awards, 44 equipment grants, 38 seed grants, and 56 standard strengthening awards (NRI Annual Report, 1999).

NRI strengthening grants have had a major effect on the careers and productivity of faculty who otherwise would not have federal grant support. That is reflected by results of a survey conducted for USDA by the Oak Ridge

Institute for Science and Education (ORISE unpublished data). The results of the survey are presented in table 4-2.

The contributions of standard strengthening grants or equipment grants to institutional teaching and research programs include the development of new courses; the initiation of additional research programs or emphases; the acquisition of complementary equipment; the establishment of new research centers and laboratories; an increase in the quality of teaching related to food, fiber, and natural resources; an increase in the quality of research related to food, fiber, and natural resources; and an improved research environment.

Enhancing Public Understanding

The NRI maintains an Internet site for distribution of information resulting from its activities. It also distributes *Research Highlights*, a newsletter about research results of NRI-sponsored research that has been published in scientific journals. The newsletter was cited favorably by the House Committee on Science (1998). The congressional report also considered the NRI Internet site as a model for other federal agencies to make their results more readily available to Congress and the public. The NRI staff's success in organizing these vehicles is in contrast with the organization's limited resources.

TABLE 4-2 Skills and Items Affected by Acquired Grant

Aspect or Item Affected	Type of Grant, %			
	Equipment	RCEA[a]	Seed	Strengthening Standard
Acquisition of new skills and knowledge	84	100	93	91
New professional ties and linkages	61	90	73	76
Increase in number. of professional presentations	38	90	72	74
Increase in number. of scientific publications	56	70	73	77
Submission of proposals for other research funding	89	80	73	69
Improved continuity of funding	50	30	41	51
Inventions	2	10	3	4
Patents, copyrights, or licensing agreements	4	0	1	5
Professional advancement in position, rank, or salary	32	20	38	37
Recruitment or retention of students	57	30	50	51

[a]RCEA = research career enhancement awards.

Source: Strengthening Award Program Assessment Results, ORISE 1997; unpublished data.

COMPLEMENTARITY

Complementarity refers to the degree to which NRI activities complement similar activities conducted by others in industry and in other federal and state agencies and programs. Complementarity also includes the extent to which the NRI reaches out to other federal and state agencies, academe, and the private sector. Complementarity—rather than isolation or duplication—is necessary and desirable for adding value to research conducted within single programs, for ensuring that overall research funds are used efficiently, and for providing a clearer path between basic research and its application. The committee provides here a brief overview of some of the research programs in USDA, other federal and state agencies and programs, and industry that complement the NRI. The research programs described should be viewed as illustrative examples, not as a comprehensive compilation of all complementary programs in the federal, state, and private sectors, which was not possible given the time and funding constraints of this study.

Other USDA-Funded Research Programs

USDA supports research on food, fiber, and natural resources through a number of research programs—inside and outside USDA—that complement the NRI. For example, three USDA agencies conduct intramural research on different aspects of USDA's mission: the Agricultural Research Service (ARS), the Economic Research Service (ERS), and the US Forest Service. USDA also provides formula funds to support research at state agricultural experiment stations, special grants to support targeted research initiatives specified by Congress, and a small amount of funds for other forms of competitive grants.

Intramural Research

Intramural support provides stable funding for long-term research activities that are central to the missions of the agencies. In FY 1998, funding for intramural research totaled $982 million, or 52% of the USDA research budget.

Agricultural Research Service. ARS conducts basic and applied research, some of it targeted at helping USDA agencies resolve scientific and technical issues that arise as they fulfill their program responsibilities. Its research is in three national programs: Animal Production, Product Value, and Safety; Natural Resources and Sustainable Agricultural Systems; and Crop Production, Product Value, and Safety. The Animal Production, Product Value, and Safety Program conducts multidisciplinary research to solve problems that threaten the security, safety, and productivity of the US food and fiber system and those arising from the interaction between animal and crop production and sustainable food and fiber systems. The Natural Resources and Sustainable Agricultural Systems Program conducts multidisciplinary research to solve problems arising from the interaction between food and fiber production and the environment.

The Crop Production, Product Value, and Safety Program conducts multidisciplinary research to solve problems that threaten the security, safety, and productivity of US food and fiber system. In carrying out its responsibilities, ARS works closely with other federal research programs and with USDA's mission agencies that rely on technology and science to carry out their program responsibilities.

Economic Research Service. ERS provides economic and other social-science information and analysis for improving the performance of agriculture and improving rural America. It collects and maintains a number of historical data series on farm type, size, and number; production and input levels; trade; effects of farm policy; and socioeconomic characteristics of rural areas of the United States. ERS also provides key statistical and analytic support to the executive and legislative branches of the federal government.

US Forest Service. The US Forest Service conducts basic and applied research on the nation's forests and on technologies useful in the manufacture of pulp and wood-based products. Research issues examined by the Forest Service include the effects of climate change on forest productivity, the behavior of fires and ecosystem response to catastrophic fires, the effects of forestry on water quality and wildlife, and methods to increase productivity through improved management.

Formula Funds: State Agricultural Experiment Stations (SAESs) and Cooperative Extension Services.

Formula funds provide matching dollars to the SAESs, which usually use these allocations for applied, state-specific research. Formula funds provide valuable flexibility by which experiment stations can respond to emerging problems and research issues. Formula funds are often used to support long-term research programs at the nation's land grant universities. In FY 1998, these funds—distributed by a formula based on the size of the farm and rural population in individual states—amounted to $516 million, or about 27% of the USDA research budget.

Special Grants

Special grants are targeted to shorter-term research needs identified by Congress. These grants are usually a minor part of the USDA research budget, but occasionally large initiatives emerge from USDA allocation bills. Recent large initiatives include the FY 1998 appropriation for plant-genome work and two initiatives included in the Agricultural Research, Education, and Reform Act of 1998. Smaller initiatives (41 in FY 2000) are often the size of individual NRI research grants (about $200,000). In some cases, funding for special grants comes at the direct expense of the NRI budget (personal communication, NRI staff, 1998). In FY 1998, special grants accounted for $169 million, or about 9% of the USDA research base.

Initiative for Future Agriculture and Food Systems. The purpose of this initiative is to support research, extension, and education activities targeted to the following research areas mandated by Congress in the Agricultural Research, Education, and Reform Act of 1998: agricultural genomics; agricultural biotechnology; food safety, food technologies, and human nutrition; new uses for agricultural products; natural-resources management, including precision agriculture; and farm efficiency and profitability. The initiative will give high priority to proposals that successfully integrate research, extension, and education or address the concerns of small and medium producers and land managers (especially in natural-resources management and farm efficiency and profitability). The goal of the initiative is to award large grants to multistate, multi-institutional, and multidisciplinary projects. The initiative has a total budget of $120 million for FY 2000.

Integrated Research, Education, and Extension Competitive Grants Program. The purpose of this program is to support integrated, multifunctional agricultural research, extension, and education activities mandated by Congress in section 406 of the Agricultural Research, Extension, and Education Reform Act of 1998. The act specified that research grants be awarded, subject to availability of appropriations, on a competitive basis to colleges and universities (as defined in section 1404 of the National Agricultural Research, Extension, and Teaching Policy Act of 1977). It specified that grants be awarded to address priorities in the US food and fiber system that involve integrated research, education, and extension activities as determined by the secretary of agriculture in consultation with the National Agricultural Research, Extension, Education, and Economics Advisory Board. Funded in FY 2000 were water quality ($13 million), food safety ($15 million), pesticide impact assessment ($4.54 million), Crops at Risk from Food Quality and Protection Act (FQPA) implementation ($1 million), FQPA Risk Mitigation Program for Major Crop Systems ($4 million), and Methyl Bromide Transition Program ($2 million), for a total program budget of $39.5 million.

Other Competitive Grants in USDA

The competitive research grants component of USDA includes three operating programs in addition to the NRI: the Office of Extramural Programs, the Small Business Innovation Research program, and the Biotechnology Risk Assessment Program. Each of those is administered by its own director. The NRI is by far the largest competitive grants program in USDA, accounting for 87% of the competitive grants awarded by USDA in FY 1998.

The Office of Extramural Programs (OEP) provides leadership and guidance in the management of the federal assistance programs related to research, education, and extension activities supported by the Cooperative State Research, Education, and Extension Service (CSREES). OEP is responsible for the execution, administration, and payments of CSREES formula funds, grants, cooperative agreements, special projects, and other federal assistance instruments to further the USDA mission.

The Small Business Innovation Research (SBIR) program in USDA makes grants to qualified small businesses to support research to develop advanced concepts related to scientific problems and opportunities in agriculture that could lead to public benefit. Objectives of the SBIR program are to stimulate technologic innovations in the private sector, strengthen the role of small businesses in meeting federal research and development needs, increase private-sector commercialization of innovations derived from USDA-supported research and development efforts, and foster and encourage participation by female-owned and socially and economically disadvantaged small business firms in technologic innovations.

The purpose of the Biotechnology Risk Assessment Research Grants Program is to assist federal regulatory agencies in making science-based decisions about the safety of introducing into the environment genetically modified organisms, including plants, fungi, bacteria, viruses, arthropods, fish, birds, mammals, and other animals. The program accomplishes its purpose by providing scientific information derived from the risk-assessment research that it funds. Research proposals submitted to this program must address risk assessment, not risk management.

Complementarity of NRI within USDA

The non-NRI research activities funded by USDA illustrate two important points about the complementarity of NRI research with respect to other USDA-funded research. First, the NRI differs from most of the research funded by USDA in that it supports researchers outside USDA on the basis of a competitive, merit-based peer-review process. Most of the roughly $1.6 billion that USDA spends on research through non-NRI programs is distributed noncompetitively through intramural research grants to USDA staff, formula funds to state agricultural stations, and special grants to states for targeted initiatives and direct grants. This allocation system does not in itself necessarily reduce the quality or relevance of the research it supports, but it runs counter to practices at the National Institutes of Health and the National Science Foundation and to the general direction of most federal research practices for assessing research quality and relevance. Second, although ARS focuses more on applied research than the NRI, there could be some overlap in the research conducted in-house through ARS and that funded through the NRI. The committee concludes that the NRI complements other USDA activities but that more emphasis should be placed on coordinating ARS and NRI agendas.

Other Federal Programs

National Science Foundation

A number of entities within the NSF Biology Directorate involve research related to food, fiber, and natural resources, including the Division of Environmental Biology (DEB), the Cluster for Systematic and Population Biology, the Cluster for Ecological Studies, the Division of Integrative Biology and Neuroscience, the Division of Molecular and Cellular Biosciences (MCB), and the Division of Plant Genome Research.

DEB supports fundamental research on the origins, functions, relationships, interactions, and evolutionary history of populations, species, communities, and ecosystems. Scientific emphases of DEB include biodiversity, molecular genetic and genomic evolution, mesoscale ecology, computational biology (including modeling), conservation biology, global change, and restoration ecology.

The Cluster for Systematic and Population Biology supports research on the patterns and causes of diversity within and among populations and species. Research projects involve any group of organisms—including terrestrial, freshwater, and marine taxa—and range in subject from microorganisms to multicellular plants, animals, and fungi. Research areas are arranged in three main groups: population biology, systematic biology, and biotic surveys and inventories.

The Cluster for Ecological Studies supports research on natural and managed ecologic systems, primarily in terrestrial, wetland, and freshwater habitats. Research areas include experimental, theoretical, and modeling studies on the structure and function of complex biotic-abiotic associations and the coupling of small-scale systems to each other and to large-scale systems. They are arranged in four groups: ecosystem studies, ecology, long-term ecological research (LTER), and long-term research in environmental biology.

The Division of Integrative Biology and Neuroscience supports research aimed at understanding the living organism—plant, animal, microorganism—as a unit of biologic organization. Such research encompasses the mechanisms by which plants and animals develop, grow, reproduce, regulate their physiological activity, and respond to their environment; the integration of molecular, subcellular, cellular, and functional genomic approaches to understanding the development, functioning, and behavior of organisms in the laboratory and in natural settings; all aspects of the nervous system, including its structure, function, development, and integration with the physiologic and behavioral systems affected by it; factors influencing the behavior of animals in the laboratory and in the field; whole-organism approaches to physiologic ecology; and the form and function of organisms in view of their evolution and environmental interactions.

MCB supports research and related activities that contribute to a fundamental understanding of life processes at the molecular, subcellular, and cellular levels. Investigator-initiated research proposals are considered in biomolecular structure and function, biomolecular processes, cell biology, and genetics. Biodiversity and biotechnology are major focal points of MCB.

The Division of Plant Genome Research was initiated in FY 1998 as part of a national plant genome research initiative established by the Office of Science and Technology Policy. The long-term goal of this program is to understand the structure, organization, and function of plant genomes important to agriculture, the environment, energy, and health. The program supports research on plant genomics and aims to accelerate the acquisition and use of new knowledge and innovative approaches to elucidate fundamental biologic processes in plants.

The Hydrologic Sciences Program in the Earth Sciences Division of NSF's Geosciences Directorate has some potential overlap with the water-quality research supported by the NRI. The program supports fundamental research on continental water processes and the global water balance. Research on the former focuses on the physical and chemical processes characterizing or driven by the cycling of continental water at all scales and on biologic processes that interact with the water cycle. Research on the latter focuses on the spatial and temporal characteristics of the water balance in the atmosphere, oceans, and continents.

National Institutes of Health

The National Institute of Environmental Health Sciences (NIEHS), one of 25 institutes and centers of NIH, supports research related to three interactive elements that play central roles in human health and disease: environmental factors, individual susceptibility, and age. The NIEHS mission is to reduce the burden of human illness and dysfunction with environmental causes by understanding each of those elements and how they interrelate. NIEHS achieves its mission through multidisciplinary biomedical research programs, prevention and intervention efforts, and communication strategies that encompass training, education, technology transfer, and community outreach. Although NIEHS covers a wide variety of issues, one research area directly related to the NRI's mission is pollution related to food and fiber production. NIEHS research in this area focuses on health effects of chemicals used in food and fiber production at high concentrations and of natural materials involved in food and fiber production (such as grain dust).

Department of Energy

Two parts of the Department of Energy (DOE) research program that are related closely to food, fiber, and natural resources are the Division of Energy Biosciences in the Office of Basic Energy Sciences (BES) and the Office of Biological and Environmental Research (BER). Both programs are in DOE's Office of Science.

The BES Division of Energy Biosciences supports fundamental research needed to develop future biotechnologies related to energy. The supported research focuses on the biologic mechanisms occurring in plants and microorganisms that could serve as renewable resources for fuel and other fossil-resource substitutes, as vehicles to restore previously disrupted environmental sites, and as potential components of industrial processes to produce new products and chemicals in an environmentally benign manner. The division supports research programs in four main areas: plant science, fermentation microbiology, extremophilic organisms, and biomaterials and biocatalysis.

The mission of BER is to develop the knowledge needed to identify, understand, and anticipate the long-term health and environmental consequences of energy production, development, and use. The mission is carried out through support of peer-reviewed research at DOE national laboratories, universities, and private institutions. Two BER divisions, the Life Sciences Division and the Environmental Sciences Division, could support some research related to food, fiber, and natural resources.

The BER Life Sciences Division manages a diverse portfolio of research to develop fundamental biologic information and to advance technology in support of DOE missions in biology, medicine, and the environment. Specific research areas include human genomics; ethical, legal, and social implications of genome research; structural-biology; model-organisms; microbial genome; and low-dose radiation.

The BER Environmental Sciences Division funds basic research in environmental processes, global change, and other subjects. Global-change research activities related to the nation's food and fiber system include studies to quantify sources and sinks of energy-related greenhouse gases (especially carbon dioxide) and studies to improve the scientific basis for assessing the potential consequences of climatic changes. The latter include the potential ecologic, social, and economic implications of human-induced climatic changes caused by increases in greenhouse gases in the atmosphere and the benefits and costs of alternative response options.

Interagency Programs

The NRI does not usually share funding of individual projects with other agencies but does attempt to cooperate in programs that further the cause of food, fiber, and natural-resources research. Joint programs that bring together several agencies to focus on common goals use available funds effectively. Such programs also provide visibility to neglected research areas, attract scientists into new areas, and focus attention on topics of broad interest. A prime example of interagency cooperation is the *Arabidopsis thaliana* Genome Sequencing Project, offered to researchers as a competition several times since its development from the 1990 "Joint NSF, NIH, USDA, and DOE Agreement on Cooperation in Support of *Arabidopsis thaliana* Genomic Analysis".

An example of cross-agency cooperation is the Joint Program on Collaborative Research in Plant Biology (offered via a DOE-NSF-USDA partnership), whose goals are to "foster the development of creative scientists trained in interdisciplinary research, to stimulate interest in research topics that need more attention, and to utilize the available funds for plant science research in the most effective manner".

More recent cooperative programs are the Terrestrial Ecology and Global Change Program (involving NSF, DOE, the National Aeronautics and Space Administration, USDA, and the National Oceanic and Atmospheric Administration) and the Opportunities in Metabolic Engineering Program (involving USDA, the Department of Commerce, the Department of Defense, DOE, and NSF). In each of those programs, proposals are recommended for funding by a single interagency peer-review panel; individual agencies then sponsor the proposals most relevant to their mission.

Although the NRI actively participates in cross-agency funding, it clearly follows rather than leads. No cross-agency program to date has been initiated by the NRI. The committee believes that the NRI suffers from being smaller than other agencies; this limits the amount of resources that it can contribute to cross-agency initiatives. Aside from memorandums of understanding and interagency coordination provided by the National Science and Technology Council (NSTC), the NRI does not have a process for establishing formal relationships with other federal agencies or for consulting and using stakeholder groups. The NRI receives advice on areas to fund from diverse sources, such as the NSTC and USDA groups, as well as from its own board of directors. However, no program is in place to consult and use natural affinity groups, such as the Farm Bureau, Grange, farm-commodity groups, agribusiness leaders, environmental interests, and the rural-development community.

Complementarity of NRI with Other Federal Programs

The descriptions of research areas funded by other federal agencies illustrate a number of important points about the complementarity of NRI research. There appears to be some overlap in the types of research that could receive funding through other federal research programs and which could be funded through the NRI. Two specific cases of overlap are the NSF Plant Genome Research Division and NIEHS research on agricultural pollution. Clearly, it is important for such overlapping programs to be coordinated with NRI research.

Aside from the specific cases discussed above, NRI research does not appear to duplicate research conducted in other federal or state agencies substantially. Most of the other federal research programs described above could potentially support research on food and fiber issues—but the breadth of those programs suggests that research on food and fiber issues is a very small component of any of them. For example, the broad mission of NSF's Division of Environmental Biology (to support fundamental research on the origins, functions, relationships, interactions, and evolutionary history of populations, species, communities, and ecosystems) could include some research on species that are important in the food and fiber system—but it would almost certainly support far more research on species outside the food and fiber system. Similarly, DOE's research on global change might include some research on the impact of global change on the food and fiber system, but other types of impacts would probably dominate DOE's research in this area.

It is unlikely that any of the other federal programs would support enough research on important food and fiber issues to constitute a coherent program of research on such issues.

Industry and the NRI

Where the science is ripe for product development and marketability, public-private partnerships will thrive. For example, there is good industry participation in genomics advisory groups. Industry scientists also participate in NRI panels, bringing their perspective to the competitive grants process. However, industry's lack of understanding of and participation in the overall NRI program was strongly expressed in the industry survey conducted by the committee (see appendix C). Reasons included the low level of funding, long response time, and concerns about the handling of proprietary information. Industry showed interest only in the ability to use NRI grants to enhance postgraduate training, particularly in the larger industries. The committee finds that industry-NRI interaction is well below what might be fruitfully pursued.

SUMMARY FINDINGS

Scientific Objectives

- The NRI program is credited for making important contributions to fundamental and applied research.
- The distinction between basic and applied research often is unclear in the food, fiber, and natural-resources sector. Such research might be thought of as a continuum with short-, medium-, and long-term objectives identified in any research area.
- A major emphasis of the NRI should continue to be the support of long-term, high-risk research with potential long-term payoffs. Much of this research would be classified as "fundamental" research in the traditional use of this term.
- The NRI appears to have delivered on its congressional mandate that at least 30% of its funds be devoted to multidisciplinary research.

Training and Education

- Training and education of graduate students and postdoctoral researchers attributable to the NRI program are substantial. Even with the small amounts and short duration of grants, training may be the major use of NRI funds among university researchers.

• Strengthening grants provided by the NRI program have had a major impact on the careers and productivity of faculty who otherwise would not receive federal grant support.

• The NRI staff has been successful in organizing several vehicles to promote public understanding of research in the food, fiber, and natural-resources area, particularly in view of the organization's limited resources.

Complementarity

• NRI complements other USDA activities, but more emphasis might be placed on coordinating ARS and NRI agendas.

• NRI does not duplicate other federal research efforts.

• Although the NRI actively participates in cross-agency funding opportunities to ensure complementarity of research efforts, it follows rather than leads in these efforts. The NRI suffers from being smaller than other agencies; this limits the amount of resources it can contribute to interagency initiatives.

• No process exists for establishing formal relationships with other federal agencies or for consulting and using stakeholder groups.

• Industry-NRI interaction is well below what might be fruitfully pursued.

5

Priorities and Priority-Setting at the NRI

A number of studies have proposed more-rigorous procedures for setting federal research priorities (OTA, 1991; NRC, 1995; McGeary and Smith, 1996). Priority-setting challenges all federal research organizations because, at some level, it forces a ranking of research investments, and this can galvanize criticism from researchers whose projects are not among those most highly ranked (OTA, 1991). Even so, in all times, especially in times of decreasing or flat research budgets, priority-setting is an essential tool for federal decision-makers, who must make choices among highly ranked projects and programs.

Effective priority-setting also can be used by a research organization as a way to "market" its research programs to those who make funding-allocation decisions. A systematic and rigorous process that identifies major gaps in knowledge, estimates how research could close these gaps, and anticipates the long-term benefits of applying the new knowledge can be effective in convincing decision-makers to increase investments in some fields of research (McGeary and Smith, 1996). A rigorous priority-setting process also can be used to establish metrics useful in the context of the 1993 Government Performance Results Act (GPRA) (Kostoff, 1997).

This chapter presents the committee's analysis of priority-setting at the NRI and suggestions for improving it. Throughout the chapter, the committee attempts to distinguish between the NRI priority-setting *process* (the procedures used to arrive at research priorities) and its research priorities themselves. The chapter begins with brief descriptions of the US Department of Agriculture (USDA) and NRI priority-setting processes and follows with the committee's analysis of the NRI process. The committee then analyzes the NRI's research priorities and offers its own suggestions for research of emerging importance.

PROCESS

Overview of USDA Priority-Setting Process

Comprehensive planning and priority-setting have characterized the USDA Experiment Station System since the early 1980s, with the Joint Council on Agricultural Research, Education, and Extension serving as a mechanism for coordination of participating federal and state partners. The planning and priority-setting process was modified during an agencywide restructuring of USDA in 1995, which created a new USDA National Advisory Committee. The Advisory Committee's role includes the responsibility to oversee and facilitate priority-setting in research and education. The involvement of stakeholders (including research users) that had characterized the priority-setting processes before 1995 was formalized and enhanced by the restructuring legislation.

In response to USDA restructuring and GPRA, the Research, Education, and Economics (REE) Mission Area of USDA established five goals toward which the planning of research and education programs will be directed and against which program performance will be measured. The goals are intended to bring together the interests of stakeholders in setting priorities by striving for

- An agricultural production system that is highly competitive in the global economy.
- A safe and secure food and fiber system.
- A healthy, well-nourished population.
- Greater harmony between agriculture and the environment.
- Enhanced economic opportunity and quality of life for Americans.

Those broad goals were established within the context of food, fiber, and natural-resources research and education. The land grant university colleges of agriculture and other research institutions and agencies are linked to the goals through various funding mechanisms administered by the Cooperative State Research, Education, and Extension Service, one of the four agencies that make up the USDA REE Mission Area.

NRI Priority-Setting Process

Priority-setting at the NRI is broadly shaped by the legislative authority that established the program. The original legislation established six broad NRI divisions.[1] Within the divisions, NRI scientific staff (primarily the division directors) play an important role in setting priorities among research needs. NRI staff rely on a variety of mechanisms for receiving external input to help shape priorities. For example, some attend scientific and professional meetings to gain understanding of current scientific trends and emerging research issues. Some use periodic input from science forums, user workshops, and communication with other federal agencies. Additional input can be sought from various research consortia, policy groups, and trade organizations and from the policy committees of the Experiment Station System.

The NRI scientific staff, the NRI chief scientist, and agency administrators are responsible for assimilating input from many diverse groups into an annual program description designed to solicit the best possible research proposals. In consultation with the NRI chief scientist, division directors, and deputy administrator, NRI scientific staff recommend changes or modifications in existing programs annually. Recommendations for the consolidation of programs or the creation of new research areas within the six divisions may be prepared by the chief scientist, division directors, program directors, or the deputy administrator. Final consensus recommendations emerge from a process chaired by the chief scientist and are presented to the NRI Board of Directors for approval.

Researchers also play an important role in the priority-setting process through the issues addressed in the proposals they choose to submit. Similarly, the peer-review panels contribute to priority-setting through their funding recommendations for proposals judged most worthy with respect to relevance and scientific merit. In this manner, the processes of planning, priority-setting, and accountability are inherently linked. NRI funding also helps to build future research capacities in high-priority areas through the support of new scientists, graduate students, and postdoctoral appointments.

Analysis of NRI Priority-Setting Process

The NRI staff and leadership have made substantial efforts and invested much intellectual capital to orient the annual program descriptions and requests for proposals to major, emerging issues that could not necessarily be clearly forseen when the authorizing statute was established. At the same time, however, parts of the priority-setting process used by the NRI staff seem unstructured, appear to be unevenly administered across NRI divisions, and are not explicitly linked to the goals and other strategic planning elements of the REE Mission Area. For example, although some NRI divisions hold user workshops regularly to solicit input on research priorities, other divisions have no discernable or regular mechanism of external input. The committee found that in some cases, changes in program areas and priorities appear to have

---

[1]Two other major research areas, Agricultural Systems (established in FY 1994) and Pest Biology and Management (established in FY 1995), are managed outside the six original divisions, and are essentially at the division level.

occurred primarily in response to the urging of vocal stakeholders rather than as the result of a deliberative priority-setting process. The committee also found that mechanisms are not well established to evaluate the effectiveness of NRI-funded research as time passes and progress occurs or to delineate how key research outcomes correlate with guiding research goals. The priorities of the NRI do not appear to be linked closely with the priorities of the Agricultural Research Service (ARS) and Economic Research Service (ERS), perhaps because the potential cross-functional nature of present research programs is not fully appreciated in either the ARS or the NRI administration.

The committee believes that these issues need to be addressed through revisions of the NRI priority-setting process. The process of setting priorities should be a continuing activity that promotes a view of the future in order to anticipate emerging research issues and to ensure adequate and continuing resources rather than a one-time effort. A successful priority-setting process should recognize the effects of research accomplishments and push the transfer of the results into practice.

A clear process for setting priorities, combined with transparent communication of the resulting priorities, could demonstrate that USDA is exercising leadership in the husbanding of scarce resources to solve major food, fiber, and natural-resources problems in an era of tight federal budgets and public accountability. The agency will then also serve its stakeholders by ensuring that research programs are in place to address major issues before they become crises.

NRI RESEARCH PRIORITIES

Because the NRI does not have a formal strategic plan, the committee has ascertained the NRI research priorities on the basis of its analysis of the funding history of the six main divisions and the numbers and types of research programs that have been supported since the NRI was initiated in 1991.

Funding History of the NRI's Six Divisions

Congress established the following six divisions when it authorized the NRI in the Food, Agriculture, Conservation and Trade Act of 1990:

- Natural Resources and the Environment.
- Nutrition, Food Quality, and Health.
- Animal Systems.
- Plant Systems.
- Engineering, New Products, and Processes.
- Markets, Trade, and Policy.

Competitive grants are awarded in those six major research areas to support basic and applied research that focuses on both national and regional research needs (and methods to transfer such research to on-farm or in-market practice).

A list of high-priority research areas mandated by Congress in 1990 is provided in table 5-1 (see also appendix A). Table 5-2 lists the major stakeholders for each of the six divisions. Funding histories for the divisions and some of the additional targeted programs from 1991 to 1997 are provided in figure 5-1 and summarized briefly below.

The funding history of the NRI and its six congressionally-mandated divisions is complicated. Awards have been made in as many as 26 programs—within a division, in multiple divisions, or mostly outside the divisions. The names of divisions have been altered over time to reflect the changes in their program areas. Awards are also made to special initiatives and "earmarks"—for example, Binational Agricultural Research and Development (BARD)—that lie wholly outside the six divisions but are funded by the NRI. The NRI's lack of a standardized grant-tracking and budgeting system, however, makes accurate reconstruction of the funding history of the divisions nearly impossible. Lack of such a system is a shortcoming in the NRI infrastructure. Appendix I outlines the details of a standardized tracking system that would be beneficial both for tracking of outcomes and for making the NRI's programs more transparent to stakeholders. The retrospective analysis provided here relies heavily on budget information given in the publicly available NRI annual reports. The reconstruction of the funding history is an approximation of the actual dollars awarded through the program since 1991.

The NRI Plants Division was created from the Plant Science Program and the Biotechnology Program, both of which existed in the USDA Competitive Grants Research Office (CRGO) before 1987. The Plants Division maintained the largest and most constant funding level between 1991 and 1997, ranging from $33.2 million to $37.8 million from 1991 to 1994. It has supported as many as 11 program areas simultaneously and now supports nine program areas. The apparent decline in funding to $21 million in 1995 was due to the excision of Pest Biology and Management from the Plants Division (figure 5-1) and into a separate free-standing major research area (see below). The Plants Division also experienced nearly continuous funding of individual programs during the 7-year period—this is generally not the case for other divisions. The actual program foci and their names in the Plants Division have probably changed less than those in any other NRI division.

Programs in the Animals Division were based on pre-existing programs in animal-science research under the auspices of CRGO. Those programs underwent a major reorganization with the advent of the NRI in 1991. Funding in this division is second only to that in the Plants Division and has been less consistent, ranging from $15.5 million in 1995 to $23.5 million in 1993. From 1993 to 1995, the Animals Division has supported up to five program areas. Most of the program areas have maintained constant funding support throughout the history of the NRI.

The Nutrition, Food Safety, and Health Division had a long history within CRGO before 1991. Support for human-nutrition research had existed in CRGO since 1978. The Human Nutrient Requirements Program was moved directly from CRGO into the NRI, and Congress designated funding for food-safety research in a new Food Safety Program. This division has experienced

TABLE 5-1 Summary of Congressionally-Mandated High-Priority Research Areas (1990)

Division (Research Area)	Research Programs
`Plant Systems	Plant genome structure and function Molecular and cellular biology Plant biotechnology Plant-pest Biocontrol Crop plant stress Improved nutrient qualities New food and industrial uses
Animal Systems	Aquaculture Animal reproduction, growth, disease, health—molecular Basis Nutrition Animal production and husbandry Animal well-being
Natural Resources and the Environment	Ecosystems Sustainable production Minimizing soil loss Effects of global climate change on agriculture Forestry Biodiversity
Markets, Trade, and Policy	International market share Decision-support systems Choices and applications of technology Technology assessment Rural economic development
Nutrition, Food Quality, and Health	Microbial contaminants Pesticide residues linked to human health Diet and health Bioavailability of nutrients Postharvest physiology Improved processing
Engineering, New Products, and Processes	New uses of and new products from crops, animals, byproducts, and natural resources Robotics Energy efficiency Computing Expert systems New hazards and risk assessment Water quality and management

TABLE 5-2 NRI Research Divisions and Stakeholders' Needs

Division	Stakeholders for Research Product
Plant Systems	Farmers, agricultural biotechnology companies, seed companies, ornamental- and forest-product companies, consumers
Animal Systems	Livestock producers, dairy farmers, meat packers, consumers
Natural Resources and the Environment	Consumers, farmers, food- and wood-processing plants, forest managers, environmental policy-makers, wildlife managers
Markets, Trade, and Rural Development	Farmers, commodity companies, state departments of agriculture
Nutrition, Food Quality, and Health	Domestic and international consumers, food-processing companies.
Engineering Products and Processes	Consumers, specialty-chemical companies, food-processing companies

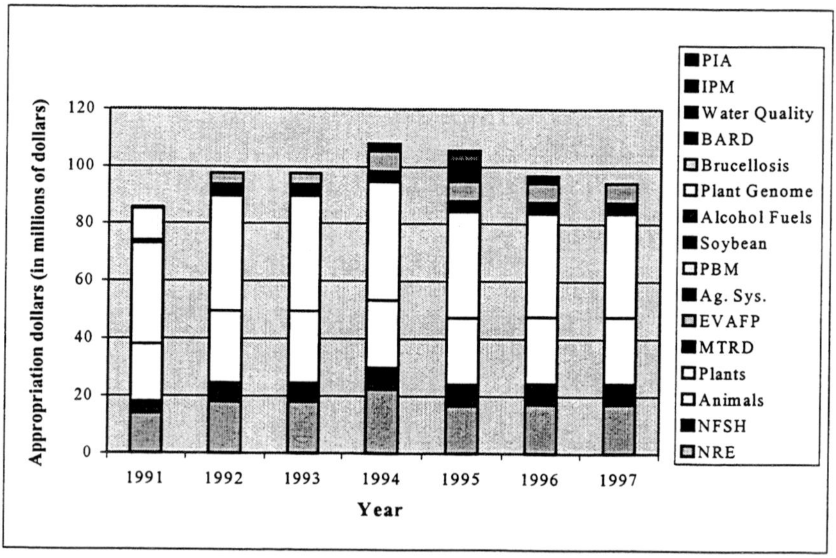

FIGURE 5-1 Congressional appropriations for NRI divisions and special initiatives, 1991–1997

NRE = Natural Resources and Environment; NFSH = Nutrition, Food Safety and Health; MTRD = Markets, Trade and Rural Development; EVAFP = Enhancing Value and Use of Agricultural and Food Products; Ag.Sys. = Agricultural Systems; PBM = Pests, Biology and Integrated Pest Management.

Source: USDA, NRI Office, 1999

considerably less funding support in the NRI than most of the others, ranging from \$3.7 million in 1991 to \$6.8 million in 1994 (figure 5-1). Only two programs (Human Nutrition and Food Safety) have maintained continuous funding.

Of all the divisions in the NRI, the Natural Resources and the Environment (NRE) Division probably has the most complicated history. Funding in the division has ranged from \$12 million in 1997 to \$20.6 million in 1994. The division has sponsored four programs, except for a reduction to three in 1998. The 1998 programs include Water Resources and Protection (WRAP), Soils and Soil Biology (SSB), and Plant Responses to the Environment. The latter began in 1985 as part of the now-defunct Biotechnology Program. It is the only consistently offered program in the NRE Division. Programs of WRAP, SSB, and Ecosystems are on a rotating basis for funding in this division.

The Enhancing Value and Use of Agricultural and Forest Products Division has had relatively stable funding ranging from \$3.8 million in 1992 to \$8.3 million in 1995. It now supports three programs.

The Markets, Trade, and Rural Development Division was first funded in 1992 and has maintained continuous funding since then for its two programs: the Markets and Trade program and the Rural Development program. It has the lowest funding of the six divisions. Funding has ranged from \$3.2 million to \$3.8 million.

Two other research areas in the NRI are outside the six divisions and are essentially at the division level themselves. Pest Biology and Management was split out from the Plants Division in 1995 and is now funded under a separate budget. Agricultural Systems was formed as a free-standing research area in 1994 to help foster interdisciplinary food and fiber research involving the natural, physical, and social sciences. All NRI divisions now help to fund Agricultural Systems.

Such interagency programs as Collaborative Research in Biology (\$135,331 in 1997), Terrestrial Ecology and Global Change (\$887,666), and the *Arabidopsis thaliana* Genome Sequencing Project (\$6.5 million) require funds from the NRI. Strengthening awards (1997 awards)—which include career enhancement awards (\$0.3 million), equipment grants (\$1 million), seed grants (\$2 million) and standard strengthening awards (\$6.7 million), in addition to postdoctoral fellowships (\$2.4 million) and new investigator awards (\$4.1 million)—also draw funds from the NRI.

Figure 5-1 shows that funding has always been unevenly allocated among NRI divisions. The figure also shows that no substantial changes in the proportion of funding allocated to each division have occurred, although the nature of food, fiber, and natural-resources research has changed since 1991. Funding allocations do not appear to have distinguished between traditional and emerging fields in food, fiber, and natural-resources research.

History of NRI Research Programs

The current portfolio of NRI research programs by division is given in figure 5-2. An important observation about the research programs in figure 5-2 is that most are organized around subdisciplines or "categories" rather than specific issues or problems that need to be solved. The development of the current 26 programs in the six main divisions can be understood in the context of the historical evolution of the NRI research agenda. The Gantt chart in figure 5-2 summarizes the initiation and consolidation of specific programs throughout the NRI's history. The chart illustrates several important points. First, several changes in program direction (such as program initiation followed by program cessation) have occurred over 4- to 6-year time frames that are shorter than would be required for the supported research to have an impact (at least 8 or 10 years). That suggests a lack of long-term strategic planning in some cases. Second, some divisions (such as the Plants and Animals Divisions) have been relatively stable programmatically since their inception, whereas other divisions (such as Natural Resources and the Environment) show a large number of program starts and stops. In general, programs with higher and more stable funding have demonstrated consistency and stability in their operations.

Analysis

The six main NRI divisions reflect potentially large, identifiable groups of stakeholders (table 5-2) and are thus a logical, first-order organizing scheme for the NRI. However, the committee believes that subdivision into the existing NRI programs solely by research "category" in the absence of an overall strategic plan is partly responsible for a lack of "critical mass" among the NRI's natural stakeholders, particularly inasmuch as the recommended increases in research funding to $550 million did not materialize. Many of the programs listed in figure 5-2 do not have strong natural constituencies among stakeholders of the food, fiber, and natural-resources system.

The mismatch between NRI programs and target stakeholders can be illustrated through an analysis of some of the entries in figure 5-2. For example, the programs listed for the Plant Systems Division do not explicitly target research on the use of genetically modified organisms that would allow a decrease in the use of chemicals in crop production—an issue that would be of great interest to both farmers and consumers. The programs in the Nutrition, Food Quality, and Health Division, do not explicitly target research on technologies to produce crops that could potentially help to prevent particular diseases; such research would be of great interest to consumers, food-processing companies, and farmers. In the Natural Resources and the Environment Division, the programs do not explicitly target research on the effects of animal-production systems on water quality—an important issue for farmers and consumers, especially in rural areas. It should be noted that the lack of explicit targeting of such issue-based research problems does not preclude NRI support

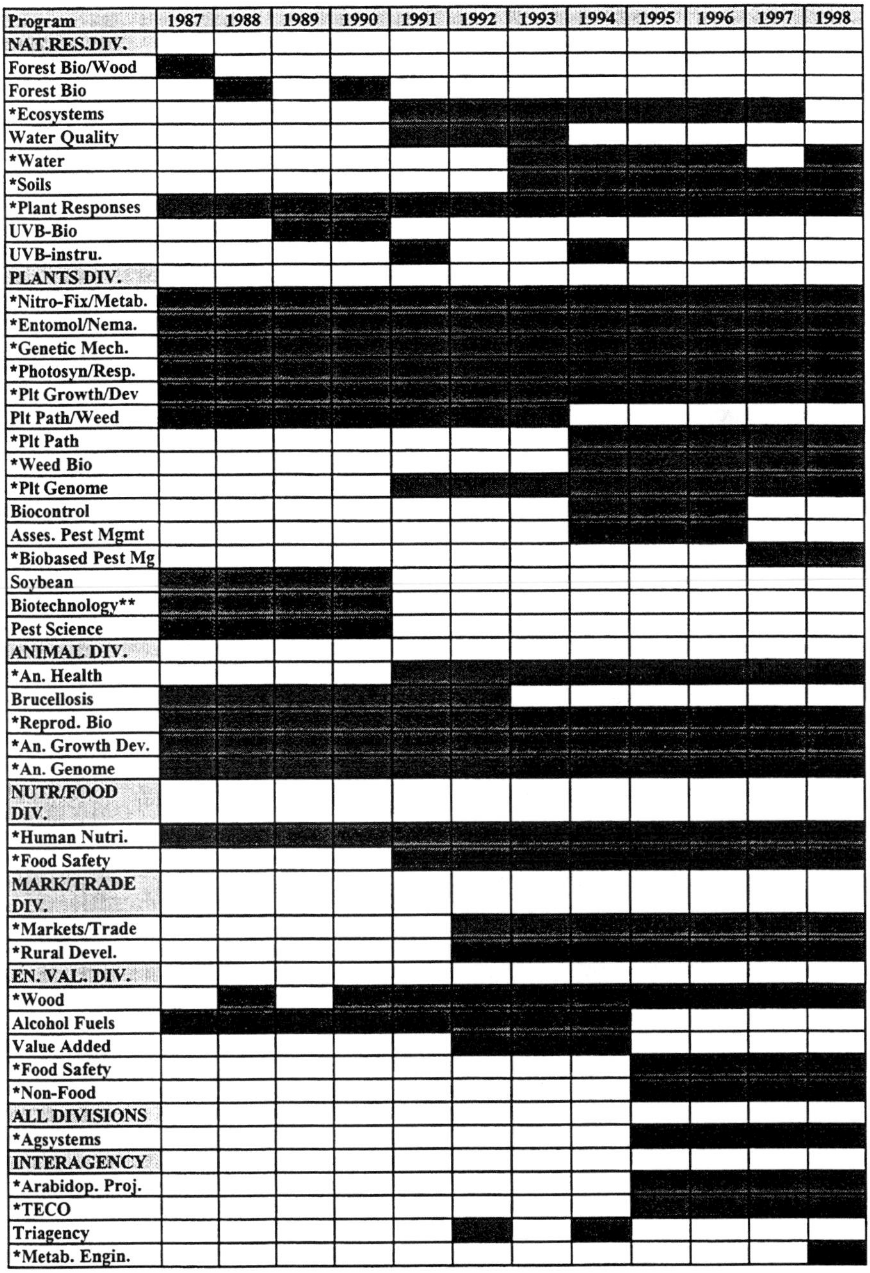

FIGURE 5-2 NRI Programs, 1987–1998 (Gantt Chart)[2]

[2]*Current program **Includes Plants and Animals Divisions.

of proposals on such topics. Rather, it simply reduces the likelihood that a substantial group of research proposals would be submitted on such issues.

The list of research issues in figure 5-2 appears to have been the result of grouping research proposals into categories representing subdisciplines rather than the result of a deliberative, priority-setting process. The committee believes that the lack of a clear perception of the logic of annual requests for proposals across the 26 programs is partly responsible for the NRI's inability to attract increased research budgets for its programs. The committee believes that a more logical priority-setting process that relates NRI programs to USDA goals and emerging issues in the food, fiber, and natural-resources system might be effective in demonstrating the importance of NRI-supported research and lead to increased research budgets.

The NRI would be much more effective if more of its programs were organized on the basis of high priority research issues or problems rather than by traditional subdisciplines. That is not to say that the NRI should focus exclusively on "applied" research—in fact, the committee sees fundamental research as an important emphasis of the NRI (see discussion of "applied" versus "fundamental" research in chapter 4)—but simply that the NRI should *organize* more of its programs around research issues or problems rather than subdisciplines. Such issue-organized research would be more easily understood by stakeholders, could be more effectively related to the rest of USDA research and to the research agendas of other federal agencies, and would encourage multidisciplinary research.

Reorganization should not lead to the elimination of all the NRI programs now organized around subdisciplines, however. The human-resource building and fundamental discipline-based research that the programs build within subdisciplines is essential and should be maintained. As McGeary and Smith (1996) point out, a healthy R&D portfolio should be driven by a mixture of disciplinary research agendas, multidisciplinary problems, agency missions, and emerging high-priority national problems.

A shift in priority-setting along the lines suggested might lead to a change in the types of research supported by the NRI. For example, past NRI funding has focused extensively on preharvest research, where 73.5% of the $87.8 million was spent in 1997. The expansion of private research funding and the increasing fundamental knowledge in preharvest (seed) technology being accumulated by industry suggest that the NRI re-examine the allocation of its resources.

The stop-start nature of some NRI funding commitments over its short history (figure 5-2) indicates that the NRI has been unable to sustain funding support for some high-risk areas with long-term payoffs—the types of research for which the NRI is ideally suited. The committee believes that there are unique opportunities for moving into long-term, fundamental research in postharvest technologies, as well as in the health and safety of the food and fiber system, that will help to add value to genetically engineered food and fiber products. The committee briefly discusses some of these areas with potential for high payoffs in the next section.

RESEARCH OPPORTUNITIES

The 1989 National Research Council report Investing in Research: A Proposal to Strengthen the Agricultural, Food, and Environmental System included a detailed list of areas for fundamental research in food, fiber, and natural resources. The committee reviewed the list and found it to be as relevant now as it was 10 years ago. The years have only added to the list of concerns. As part of its own study, the committee developed list included in appendix H and summarized in table 5-3. Although this list is not as exhaustive and does not provide as much detail as the one in the 1989 report, it is generally consistent with the 1989 conclusions.

The committee's list reflects the impacts of rapidly increasing consumer interests in health, nutritional value, and safety; the advent of bioengineering; globalization of the economy; increased awareness of environmental degradation; and the social consequences of the industrialization of the agricultural sector. In comparison with attitudes of 10 years ago, consumers and researchers alike in 2000 have heightened concern about all subjects. The committee's list is intended to be an illustrative, not comprehensive, example of how some NRI programs could be organized around research issues rather than subdisciplines. If the NRI adopted this approach, a logical way to develop the list of programmatic issues would be to have an advisory committee for each division create a similar list of emerging research issues (such advisory committees are discussed in chapter 7).

SUMMARY FINDINGS

On the basis of analysis of data submitted in various forms by NRI staff, the committee presents the following findings regarding the relationship between priority processes and funding allocations within the NRI.

Priority-Setting Process

• The priority-setting process used by NRI staff seems unstructured, is unevenly administrated across NRI divisions, and is not explicitly linked to the goals and other strategic planning elements of the REE Mission Area. For example, although some NRI divisions hold user workshops regularly to solicit input on research priorities, other divisions have no discernable or regular mechanism of external input.

• Changes in program areas and priorities appear to have occurred primarily in response to the urging of vocal stakeholders rather than as the result of a deliberative priority-setting process.

• Mechanisms are not well established to evaluate the effectiveness of NRI-funded research as time passes and progress occurs or to delineate how key research outcomes correlate with guiding research goals. The priorities of the NRI do not appear to be linked closely with the priorities of ARS and ERS, perhaps because the potential cross-functional nature of present research programs is not fully appreciated in either the ARS or the NRI administration.

TABLE 5-3 Committee's List of Emerging Research Issues

Division	Research Issues[a]
Plant Systems	Gene and genome interactions and bioinformatics Transgenic plants for improved production Mechanisms of pest-plant and plant interactions with beneficial organisms Development of the knowledge base to facilitate a new generation of biologically based materials Engineering of plant biosynthetic and metabolic pathways
Animal Systems	Gene and genome interactions and bioinformatics Functional foods and nutrient research Transgenic and cloned animals Animal reproduction Animal nutrition Animal-rangeland interactions Animal health Animal growth and development Consumable animal products Immunology Construction of novel microorganisms Gene-based pharmaceuticals and gene therapy Evolution of biologic systems
Natural Resources and the Environment	Water quality Animal-waste handling Environmental impacts Impact of biotechnologic modifications of plants and animals on the microbial ecology of reasulting food products Bioprocess engineering Biodiversity Weather and climate interactions in agricultural systems Global change and agriculture Nitrogen-use efficiency Wildlife in agricultural systems Space research

Markets, Trade, and Policy	Development of a knowledge base to prepare for biologic terrorism
	Globalization of the economy
	Identification of the economic and social consequences of environmental regulation
	Improvement of farm income and risk-management tools
	Examination of the impacts of the changing farm and agribusiness structure
	Evaluation of trade policies and barriers
	Development of effective economic and rural community development programs
	Assess how changes in consumer demand affect health, nutrition and food safety
	Analysis of economic and social impacts of consolidating research and extension programs
	Examination of current and emerging information technologies and communication systems
	Improved understanding of economic and social impacts of biotechnology
	Investment in human capital development
Nutrition, Food Quality, and Health	Research on nutrient-drug interactions
	Assessment and characterization of the impact on consumers of phytochemical substances
	New and resurgent pathogens in foods
	Pasteurization and sterilization of foods
	Identification and modification of allergens in foods
	Probiotic development
Enhancing Value and Use of Agricultural, Food, and Forest Products	Development of analytic microtechnology
	Impacts of organic farming
	Bioprocess engineering of agricultural products
	Metabolic pathway analysis and structure

See appendix H for more-detailed discussion of each issue.

Research Priorities

• Funding has been unevenly allocated among NRI divisions from the beginning. No substantial changes in the proportion of funding allocated to each division have occurred, even though the nature of food, fiber, and natural-resources research has changed since 1991. Funding allocations do not appear to have distinguished between traditional and emerging areas in the food, fiber, and natural resources research.

• The subdivision of the NRI's six main research areas into existing NRI program areas by research category in the absence of an overall strategic plan could be partly responsible for a lack of critical mass among the NRI's natural stakeholders, inasmuch as the recommended increases in research funding to $500 million did not materialize.

- A shift in priority-setting might lead to a change in the types of research supported by the NRI.
- In general, programs with higher and more stable funding have been more consistent in their operations, whereas other divisions show a large number of program starts and stops during the NRI's history.
- Several short-term changes in program direction (over 4- to 6-year time frames) have occurred in research areas that would otherwise need at least 8 or 10 years to have an impact. That suggests a lack of long-term strategic planning in some cases.
- The lack of a clear perception of the logic of annual requests for proposals across the 26 programs is partly responsible for the NRI's inability to attract increased research budgets for these programs. The committee believes that a more logical priority-setting process that relates NRI programs to USDA goals and emerging issues in the food, fiber, and natural-resources system might be effective in demonstrating the importance of NRI-supported research and lead to increased research budgets.

Overall, the process of NRI priority-setting appears to be reactive, not active. Change has come about because vocal groups advocated areas of scientific opportunity (NRI leadership and principal-investigator constituency) rather than because of clear mission focus and research strategy. Systems to relate to all constituencies regularly to share input and review mission have been ad hoc. The committee believes that there is ample room in the six congressionally mandated divisions to redefine a consistent focus for funding and to adjust that focus as the long-term priorities of the food, fiber, and natural-resources systems change.

As a major, peer-review-based research-funding mechanism, the NRI should have its programs more closely linked to the overall goals, planning, and evaluation procedures of the food, fiber, and natural-resources system. The linkage should reflect the NRI's mission relative to other funding mechanisms and programs (see "Complementarity" in chapter 4). Included in the NRI's role should be strong emphases on fundamental and multidisciplinary research and mission-linked and single-discipline approaches (see "Scientific Objectives" in chapter 4).

6

Organizational and Funding Issues

Previous chapters have summarized the committee's assessment of the quality, fairness, relevance, and responsiveness of the NRI competitive grants program; the program's priority-setting processes and research priorities; and the program's overall role and scope within the nation's research and development enterprise. Issues directly related to the NRI's organization and funding have been raised repeatedly during the committee's analysis of these subjects. For example, many respondents to the committee's survey indicated that the impact of the program, although important, has been limited by an inadequate budget and by awards that are too short, too few, and too small (see chapter 3). Similarly, the committee found (chapter 4) that the NRI is too small to take an active role in interagency research initiatives. The committee found (chapter 5) that the NRI's formal priority-setting process needs improvement and that organizational changes were warranted. And the committee found (chapters 4 and 5) that mechanisms are not well established to coordinate NRI research goals with those of complementary research organizations in the US Department of Agriculture and in other federal agencies.

Those findings suggest that organizational and funding issues play an important role in the nature and content of changes that will be required to make the NRI a more effective part of the nation's research efforts in the food, fiber, and natural-resources area. In the first half of this chapter, the committee briefly considers several organizational issues, such as the location of the NRI in USDA and issues related to its day-to-day governance. In the second half, the committee discusses important issues related to NRI funding, including the total budget for the program, the average size and length of grants, and the limit on overhead rates.

ORGANIZATIONAL ISSUES

Location in USDA

USDA's Research, Education, and Economics mission area comprises four agencies—the Agricultural Research Service (ARS), the Economics Research Service (ERS), the National Agricultural Statistics Service, and the Cooperative State Research, Education, and Extension Service (CSREES) (figure 6-1). CSREES comprises nine units, including the Competitive Research Grants and Awards Management Division (CRGAM). Other CSREES units are responsible for allocating formula funds and special grants to land grant institutions, agricultural experiment stations, and cooperative extension services. The NRI is one of four operating units in CRGAM (figure 6-2). The NRI therefore is two organizational levels below USDA's main intramural research organizations—ARS and ERS.

A previous National Research Council report (NRC, 1989) presented four criteria to evaluate the strengths and weaknesses of various options for locating an expanded competitive grants programs (the NRI) within USDA. In particular, for such a competitive grants program to be successful, the location should

- Ensure the program's openness to high-quality science and provide it with broad appeal, visibility, and stature in the scientific community.
- Provide the program director and chief scientists with direct access to key high-level policy-makers in USDA.
- Develop strong relations between the competitive grants program and the research programs of other agencies.
- Attract nationally prominent scientists and managers to positions of program leadership and to service on program advisory committees and peer-review panels.

On the basis of the analyses presented in the previous chapters and the results of the committee's survey (see appendix C), the committee believes that the location of the NRI in USDA has several shortcomings with respect to those criteria. First, the location does not provide the NRI with broad appeal, visibility, and stature in the scientific community. Second, the two organizational levels between the NRI and the under secretary for research, education, and economics (CSREES and CRGAM) can limit the access of NRI leaders to such high-level USDA policy-makers. Third, the NRI's location

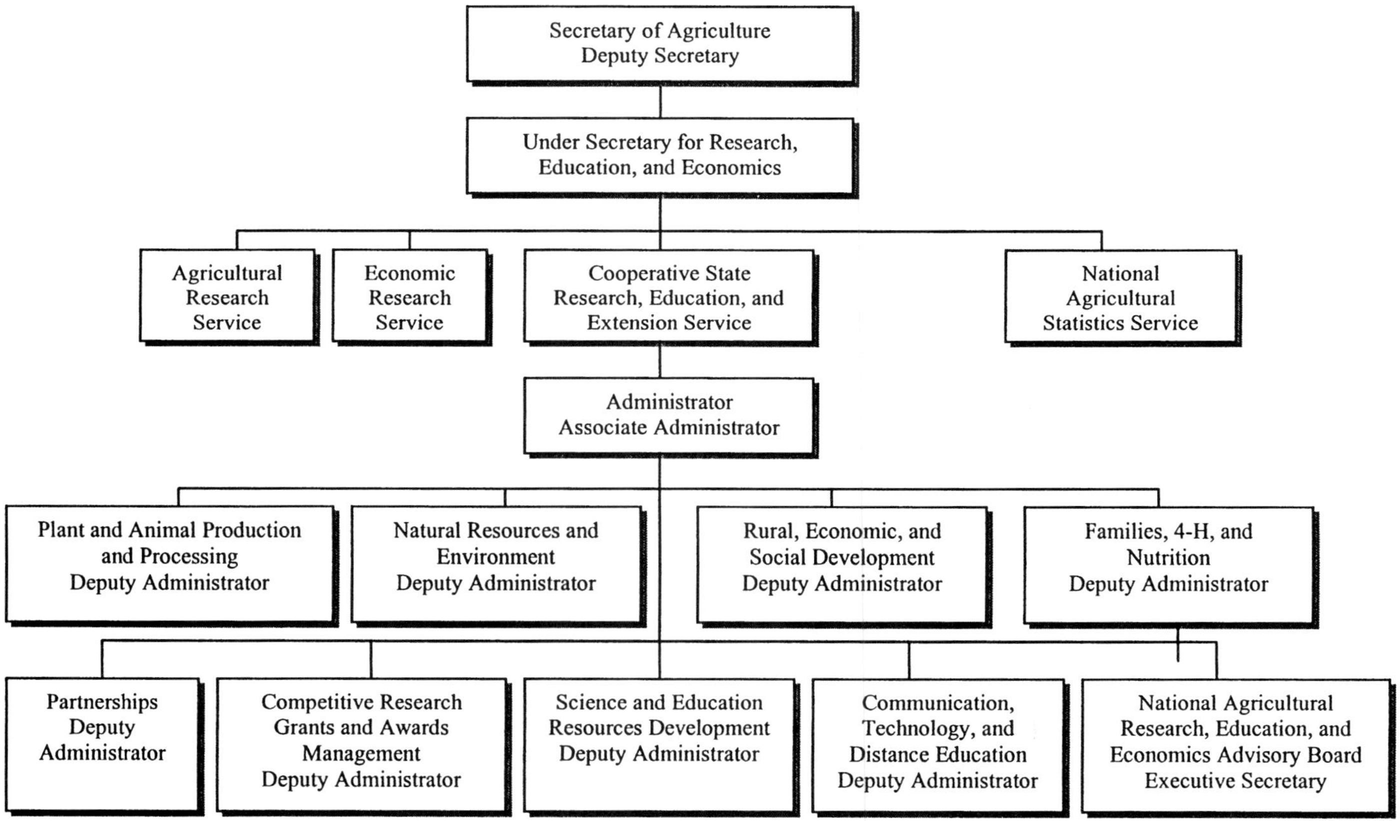

FIGURE 6-1 Organization of Research at the US Department of Agriculture

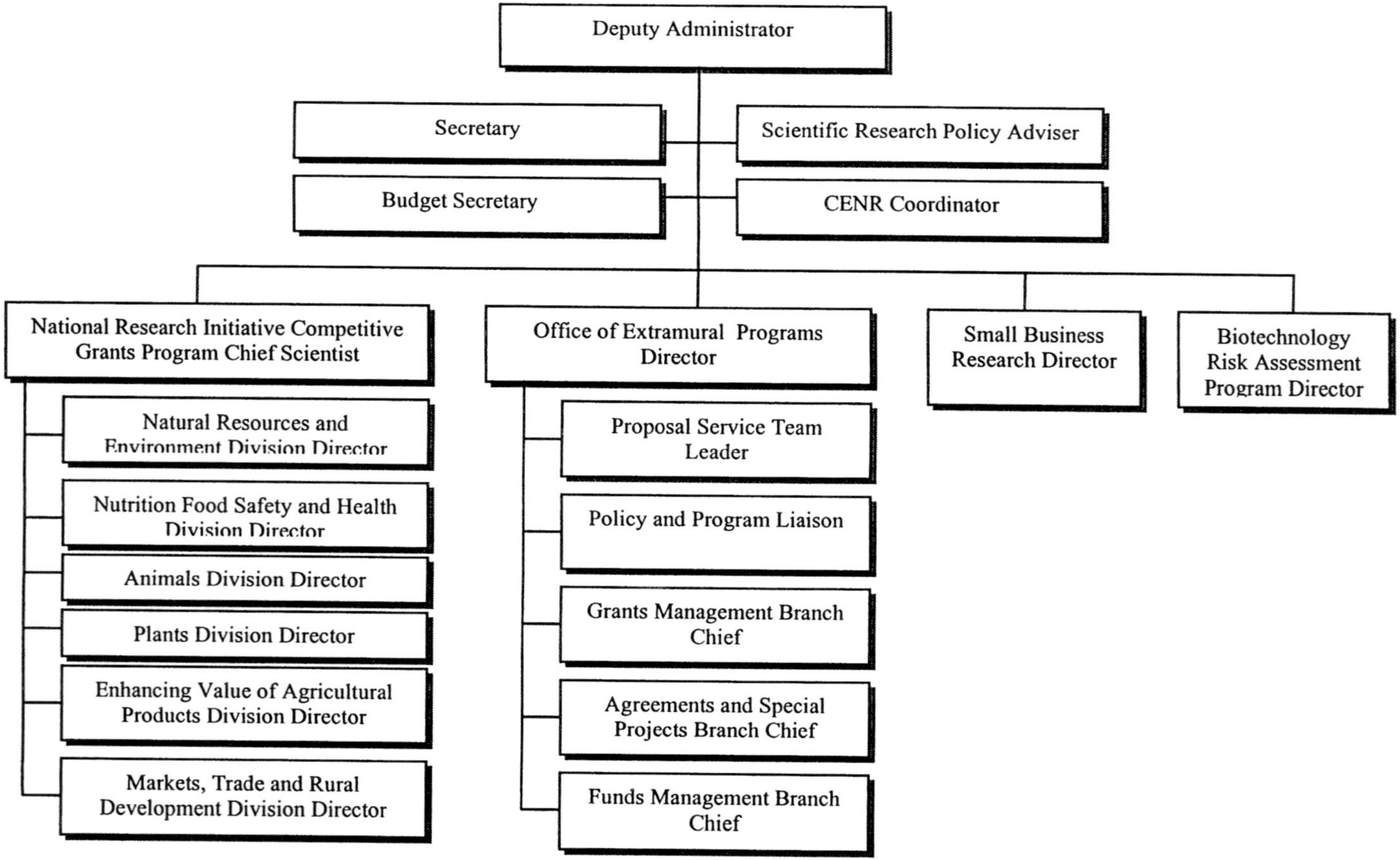

FIGURE 6-2 Competitive Research Grants and Awards Management

might be partly responsible for the tendency of the NRI to take a supportive, rather than leadership, role in interactions with other federal research agencies (see chapter 4). Finally, the location of the NRI as one component of CRGAM, rather than on an organizational level equivalent to USDA's two main research agencies, suggests that USDA and Congress place a higher priority on formula funds, special grants, and intramural research than on extramural, merit-based peer-reviewed research. The committee believes strongly that unless extramural competitive research is given the same stature organizationally in USDA that formula-funded research and intramural research receive, it might remain difficult for the NRI program to achieve its mission.

NRI GOVERNANCE

Chief Scientist

The scientific leadership for the NRI is provided by a chief scientist who functions as the director of the NRI. The chief scientist holds a part-time, apolitical, nonadvocacy position. Candidates are generally recruited from academe and serve 2-year terms. Since the program's inception, all chief scientists have been members of the National Academy of Sciences. The stated responsibilities of the chief scientist are to

- Interact regularly and directly with the under secretary for research, education, and economics, the administrator of CSREES, the deputy administrator of CRGAM, the Board of Directors of the NRI, and other administrators and staff scientists of CSREES.
- Establish policies for the NRI in consultation with NRI division and program directors and administrators listed above.
- Serve as the principal communicator for the NRI with representatives of federal and state agencies, private organizations, and special-interest, academic, professional, and commodity groups.
- Interact with division and program directors day to day.
- Oversee the peer-review process used to assess the merits of research proposals received for consideration by the NRI.
- Provide general scientific leadership responsibilities, including supervising the preparation of program descriptions and requests for proposals; publications of the NRI, such as its annual report; and *NRI Highlights*.
- Allocate NRI appropriations to the panels after merit review is completed.
- Serve as a member of the NRI Board of Directors.

In the committee's view, those responsibilities are equivalent to a full-time position. This view is shared by the four former chief scientists who were interviewed by the committee (see appendix C). It became clear to the committee that the current part-time, revolving chief scientist cannot meet the strategic-planning, priority-setting and communication needs of an effective NRI (see also chapter 5). Having a chief scientist who serves part-time hampers continuity in accountability and leadership and counters successful long-range planning and followup and consistent stakeholder involvement.

Board of Directors

The NRI Board of Directors meets regularly and determines policy for the program. The under secretary for research, education and economics chairs the Board, which also includes the administrators of CSREES, ARS, and ERS; the deputy chief for research of the US Forest Service; and the NRI chief scientist. The NRI executive officer is the deputy administrator of CRGAM (figure 6-2). The NRI Board of Directors provides administrative oversight of the NRI program and can be used to link the NRI with USDA's other research organizations.

The NRI Board of Directors is not responsible for providing guidance on scientific or technologic priorities, providing a forum for stakeholder concerns, or measuring research outcomes and the evaluation of NRI operations. The committee believes that an external advisory board of some type is necessary to fulfill these responsibilities (see discussion in chapter 7).

Organization

The NRI has six divisions. Ideally, each division has a permanent director, who oversees all operations involved in the application, review, and award processes. In 1998, three division directors were managing the six divisions (see figure 6-3). One, for example, was responsible for program areas within the Plant Systems Division and the Markets, Trade, and Rural Development Division, even though this director did not have substantive training in social science (in fact, the committee observed a general lack of social-science expertise among NRI staff in 1998). Program directors provide scientific oversight of individual research programs and, with rotating panel managers recruited from the research community, are responsible for administering NRI review panels.

Implicit in the NRI table of organization is a range of intellectual and administrative tasks that sustain the integrity of any competitive grants process. These include ensuring continuity across program areas, regulating workload in proposal handling, and determining award amounts on the basis of panel rankings of priorities. In recent years, the NRI staff has been stretched to cover those tasks, increasing the burdens of communication and timeliness on NRI staff at all levels and on the all-important scientists who serve as ad hoc reviewers and panel members.

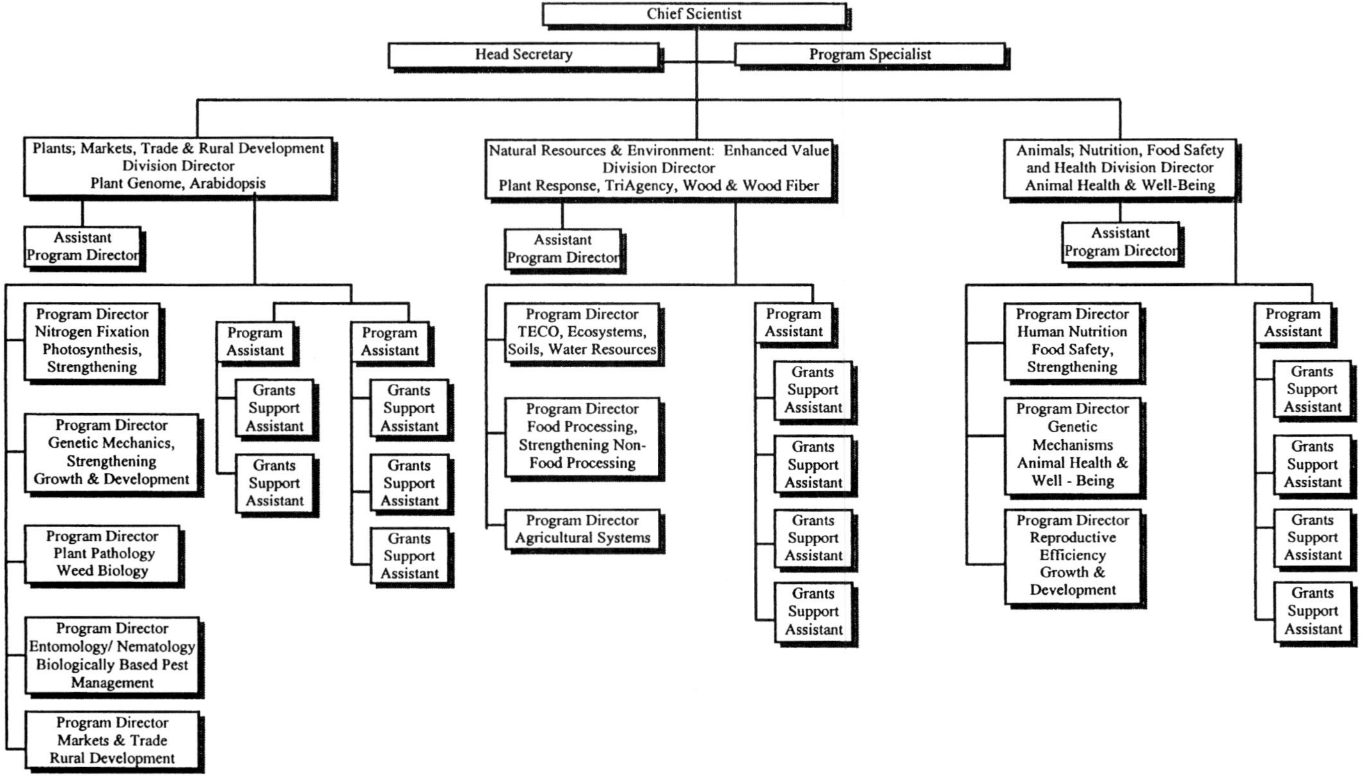

Figure 6-3 National Research Initiative Competitive Grants Program Organization

FUNDING ISSUES

Funding has been a recurring theme throughout the committee's study. Nearly all survey respondents, chief scientists, and those who testified before the committee expressed the view that the total budget for the program was inadequate and that awards were too short, too few, and too small (appendix C). The low level of funding has limited the NRI's ability to take a lead role in interagency initiatives (see chapter 4) and might have contributed to the recent decision to locate the Plant Genome Research Project in the National Science Foundation (NSF) rather than USDA. It also has led to a substantial reduction in application numbers from 1994 to 1998 (see appendix F). In the following sections, the committee briefly discusses three important components of the NRI's funding: the total budget for the program, the average size and length of grants awarded by the program, and the congressionally mandated limit on overhead rates.

Total NRI Funding

When the NRI was established in 1991, its initial funding goals were designed to ensure the growth of a dynamic research program. The enabling legislation, Public Law (PL) 101-624, authorized $150 million for 1991, $275 million for 1992, $350 million for 1993, $400 million for 1994, and $500 million for 1995 (PL 101-624, 101st Congress, Federal Register). The amount of funding appropriated, however, has never approached those optimistic goals. NRI funding fell short in the very first year, when funding was appropriated at only $73 million. Although nearly double the amount for the preceding competitive grant programs in USDA ($42.5 million in 1990), the 1991 appropriation was only about half the authorized amount. As a result, the program funded only four of the recommended program divisions (now titled Plants; Animals; Nutrition, Food Safety, and Health; and Natural Resources and Environment). Programs in the remaining two divisions (Markets, Trade, and Rural Development; and Enhancing Value and Use of Agricultural and Forest Products) and in strengthening awards (Career Enhancement, Equipment, Seed, and Standard Strengthening) were not initiated until 1992.

Despite the intended increase in NRI funding from 1991 to 1995, appropriations remained at or near the $100 million level during that period (see table 6-2). Special initiatives or "earmarks", such as the BARD Program in 1994 and 1995, cut into the NRI budget and effectively decreased the total funding available to the six original NRI divisions. The NRI budget remained flat at approximately $100 million until FY 1999, when the budget was increased to nearly $120 million (see table 6-2).

The NRI has labored under the expectation of a $500 million research portfolio, although federal budget pressures have maintained annual appropriations far below this authorized level. NRI staff testified that some researchers have expressed a reluctance to submit proposals to the NRI because

TABLE 6-1 NRI Funding Levels, 1991–1998

Year	No. Awards	Total Amount Awarded, $	Average Grant Award,[a] $	Average Grant Length, years	Average Funding, $/year
1991	590	69,204,000	NA[b]	NA[b]	52,591
1992	777	92,138,350	126,998	NA	NA[b]
1993	790	91,814,480	124,846	2.1	59,450
1994	833	96,631,441	137,256	2.35	58,407
1995	783	93,796,282	127,773	2.13	59,987
1996	739	87,801,344	125,620	2.14	58,701
1997	712	87,315,733	133,379	2.6	51,300
1998	699	88,106,761	136,065	2.2	61,848

[a]Excluding research career enhancement awards, equipment grants, and seed grants.
[b]Not available.
Source: NRI annual reports for 1991, 1992, 1993, 1994, 1995, 1996, 1997, and 1998.

of a combination of factors: modest budget sizes, low success rate, small awards, short grant duration, and the low 14% overhead cap (which many institutions will not accept). The recent increase to a 19% overhead rate is not expected to change the situation substantially.

Failure to obtain the originally proposed appropriations has stunted the development of the NRI and has challenged its effectiveness, potentially reducing the desired number of high-quality research grants with sufficient size and duration to achieve research goals. The practical result has been that a large pool of US scientists might not have been fully used in research directed to issues critical to the food, fiber, and natural-resources system. The committee concludes that inadequate funding of the NRI has significantly limited its potential and placed the program at risk.

Size and Length of Grants

As shown in table 6-1, the number of grants awarded in a single year has ranged from 590 (in 1991) to 833 (in 1994). Between 1993 and 1998, the average annual funding level has remained relatively constant at about $60,000 per year. Over the same period, the average grant length remained relatively

TABLE 6-2 History of Funding for Food, Fiber, and Natural-Resources Research in the USDA (in millions of dollars)

Year	Intramural				CREES							Total
					Formula Funds		Competitive	Special Grants		Other		
	ARS	FS	ERS	NAL	Research	Extension	Grants	Research	Extension	Research[a]	Extension	
1985	491.4	113.8	46.6	11.5	197.1	260.2	53.8	32.0	77.6	1.5	5.9	1,291.4
1986	483.2	113.6	44.1	10.8	189.0	260.2	48.8	30.2	78.9	1.6	5.5	1,265.9
1987	511.4	126.7	44.9	11.1	189.0	254.1	46.7	55.1	78.6	2.9	6.3	1,326.8
1988	544.1	132.5	48.3	12.2	201.8	260.8	45.4	51.8	80.2	4.1	16.9	1,398.2
1989	569.4	138.3	49.6	14.3	202.8	260.8	39.7	41.9	82.0	6.4	18.6	1,423.8
1990	593.3	150.9	51.0	14.7	202.8	265.1	42.5	73.1	86.4	8.2	18.2	1,505.7
1991	631.0	167.6	54.4	16.8	212.0	276.4	73.0	78.6	103.4	9.7	18.7	1,641.6
1992	668.4	180.5	58.7	17.8	220.3	288.5	97.5	87.1	110.0	10.6	20.9	1,760.2
1993	668.0	182.1	58.9	17.7	220.3	288.6	97.5	73.4	118.0	10.5	18.4	1,753.4
1994	691.6	192.5	55.3	18.3	225.9	298.1	112.2	72.9	117.4	12.1	19.1	1,815.3
1995	752.0	194.0	53.0	18.0	226.0	301.0	101.0	75.0	112.0	10.0	16.0	1,858.0
1996	704.0	178.0	53.0	19.0	217.0	296.0	94.0	75.0	107.0	11.0	12.0	1766.0
1997	697.0	180.0	53.0	20.0	250.0	327.0	94.0	80.0	71.0	57.0	28.0	1,857.0
1998	722.0	188.0	71.6	20.0	221.0	295.0	97.0	65.0	104.0	86.0	20.0	1,889.6
1999	795.0	197.0	65.0	19.0	236.0	305.0	119.0	71.0	109.0	62.0	19.0	1,997.0

[a] Includes Fund for Rural America.

Source: USDA Office of Budget Policy Analysis

constant at about 2.2 years.[1]

As discussed in chapter 4, NSF, the National Institutes of Health (NIH), and the Department of Energy (DOE) support competitive research projects in some of the same basic science and engineering fields as the NRI, that are complementary with food, fiber, and natural-resources research. The median annualized research award for NSF's Biology Directorate in FY 1998 was $90,000 per year (total costs including overhead) with an average grant duration of 2.9 years. The Biology Directorate estimates that for FY 2000, its median annualized research award will be approximately $105,400 with an average grant duration of 3.0 years (http://www.nsf.gov/bfa/bud/fy2000/00BIO.htm). Similarly, the Division of Energy Biosciences of DOE's Office of Basic Energy Sciences averaged close to $100,000 per year for grants awarded typically for a 3-year duration (http://www.er.doe.gov/production/grants/fr99_07.html). A comparison of those data with table 6-1 shows that on the average NRI research grants are much smaller and shorter than grants supporting similar types of research in NSF, NIH, and DOE. Continued underfunding of NRI research grants relative to those of other federal research agencies will tend to discourage new researchers outside the traditional food and fiber system from applying for NRI grants—one original goal of establishing the NRI (to "seek the widest possible participation of qualified scientists"). It might also cause highly qualified scientists who have received NRI support to apply for research funds from other sources and possibly to redirect their research away from issues important to the food and fiber system. The sharp decrease (over 20%) in the number of new proposals received from 1995 to 1998 (appendix F) suggest that this is occurring.[2] Such trends could lead to a decrease in the overall quality of food, fiber, and natural-resources research. The low funding levels and short grant durations and their effect on the functioning of the NRI was addressed in a 1995 Office of Technology Assessment report, *Challenges for U.S. Agricultural Research Policy* (OTA 1995):

> Thus, on the critically important issue of funding of individual awards—in terms of amount of award and duration—the program is woefully inadequate, especially in comparison to the closely related comparison programs in NSF and NIH, and little improvement has been made between earlier Competitive Research Grants program and NRI. [P. 34]

The small grants of short duration have resulted in a dwindling enthusiasm for NRI grants in the food and fiber scientific community, especially in view of the substantial administrative burden of proposal preparation. For example, one corporate scientist observed that the NRI program could actually have an adverse effect on research productivity because "the cost to the scientific enterprise nationally may exceed the funding received" owing to administrative

---

[1]FY 1997 was anomalous in that the average award length was greater and the average annual award amount lower than in other years.

[2]The increase in new applications that accompanied the budget increase in FY 1999, however, suggests that this effect can be reversed to some extent by increases in funding.

overhead of grant preparation and submission coupled with low funding levels. At very low funding rates, the effort expended by scientists in writing unsuccessful applications can exceed that of the scientists who receive research support. Some have argued that such a program is a net burden rather than an asset to the scientific community as a whole (Chubin, 1998).

Overhead Rates

When it established the NRI program in 1991, Congress imposed a 14% limit on the amount of indirect costs that can be charged as a percentage of the total award.[3] The 14% limit was replaced by a 19% limit[4] in FY 2000 as part of the Agricultural Research, Extension, and Education Reform Act of 1998. Although the increase from 14 to 19% reduces the gap between overhead rates on NRI grants and rates on grants awarded by other federal agencies, overhead rates for most academic and private-sector research institutions are significantly higher than the 19% limit currently allowed. Average overhead rates for NSF's Biology Directorate, for example, are approximately 45% of the modified total direct costs of the award—nearly double the NRI limit. The committee is not aware of any other federal merit/peer-reviewed research program with such a congressionally mandated limit on overhead rates.

Presumably, the motivation for setting such a limit was to increase the percentage of NRI research funds spent on research activities. However, such a mandated cap on overhead may have a negative effect on the NRI program because it causes some institutions (especially those from outside the traditional applicant community) to discourage their researchers from submitting proposals to the program. Because the committee did not address this issue in its survey, it was not able to estimate the magnitude of this effect on the NRI program. However, the committee is aware of one research institution that prohibits its scientists from submitting proposals to the NRI because the low overhead rates do not cover the true institutional costs associated with such research and because its auditors require consistency among all incoming grants. Other institutions discourage their researchers from submitting proposals by requiring that the researchers (or their departments) use other funds to make up the difference between mandated low overhead rates and the established rates used by other federal agencies. This is especially problematic for smaller institutions where researchers do not have the flexibility to balance low-overhead grants against other sources of unrestricted funds. These factors also may have a disproportionate impact on institutions (or departments) from outside the traditional food, fiber, and natural resources system because they do not have a historic association with the USDA, and hence, may be less willing to accept a low overhead rate that is unique to USDA-sponsored research.

---

[3]This limitation is equivalent to 0.16279 of the total direct costs of an award.
[4]This limitation is equivalent to 0.23456 of the total direct costs of an award.

SUMMARY FINDINGS

Organization

• The location of NRI as one component of CRGAM, rather than on an organizational level equivalent to USDA's two main research agencies, suggests that USDA and Congress place a higher priority on formula funds, special grants, and intramural research than on extramural, merit-based peer-reviewed competitive research. The committee believes strongly that unless extramural competitive research is given the same stature organizationally within USDA as formula-funded and intramural research receive, it might remain difficult for the NRI program to achieve its mission.

• The responsibilities of the NRI chief scientist are equivalent to a full-time position. The part-time, revolving chief scientist cannot meet the strategic-planning, priority-setting and communication needs of an effective NRI. Having a chief scientist who serves part-time hampers continuity in accountability and leadership and counters successful long-range planning and followup and consistent stakeholder involvement.

• The NRI Board of Directors provides necessary administrative oversight of the NRI program and can be used to link the NRI with USDA's other research organizations. The Board of Directors is not responsible for providing guidance on scientific or technologic priorities, providing a forum for stakeholder concerns, or measuring research outcomes and the evaluation of NRI operations. An external advisory board of some type is necessary to fulfill those responsibilities.

• In recent years, the NRI staff has been stretched to cover its responsibilities, increasing the burdens of communication and timeliness on NRI staff at all levels and on the all-important scientists who serve as ad hoc reviewers and panel members.

Funding

• Inadequate funding of the NRI has significantly limited its potential and placed the program at risk.

• NRI research grants are much smaller and shorter than grants supporting similar types of research in NSF, NIH, and DOE. Continued underfunding of NRI research grants relative to those of other federal research agencies will tend to discourage new researchers outside the traditional food and fiber system from applying for NRI grants—one original goal of establishing the NRI. It might also cause highly qualified scientists who have received NRI support to apply for research funds from other sources and possibly to redirect their research away from issues important to the food and fiber system. This could lead to a decrease in the overall quality of food, fiber, and natural-resources research.

• Congress imposed a 14% overhead limit on the NRI when it established the program in 1991. The 14% limit was replaced by a 19% overhead limit in

FY 2000. There is no clear reason why the NRI is treated differently from other federal peer-reviewed research in this regard.

7

Recommendations

Since the late 1800s, the publicly supported system of food, fiber, and natural-resources research and education in the United States has served as a model for directing scientific and financial resources to improve societal well-being. Payoffs from this research and education system have consistently been high. Moreover, the US system has been emulated successfully by many other countries. As we enter the 21st century, however, this traditional system has evolved to include a broader set of issues that can be addressed through high-quality fundamental research, technology transfer, outreach, and education.

The modern system attempts to integrate food, fiber, and natural-resources issues and increased economic opportunities to enhance the quality of life of families and communities. Fundamental research is vital to provide the depth and breadth of knowledge needed for solving societal problems and creating new opportunities to improve the quality of life. Issues high on most agendas for food, fiber, and natural-resources research include a safe, nutritious, and affordable food supply; global competitiveness; a cleaner environment; and prudent conservation of natural resources.

The committee believes that the US Department of Agriculture (USDA) National Research Initiative Competitive Grants Program (NRI), although operating well below its intended level, is a platform on which a re-energized national initiative for research in food, fiber, and natural resources can be established. Substantive research contributions have originated in the NRI. The NRI is facing operating challenges largely as a consequence of inadequate funding, which has prevented it from crossing a threshold to sustainability and growth. But after some 20 years of merit-based peer review in the USDA and the 9-year history of the NRI, the NRI is a successful template to support a substantial increase in public research in national food, fiber, and natural resources.

The NRI or an equivalent merit-based peer-reviewed research effort is needed to lead and shape our nation's response to the challenges of food, fiber, environment, energy, and a rapidly growing global population in the 21st century. A knowledge base of information and technology unprecedented in the history of the natural sciences is needed today. The committee makes the following recommendations to strengthen the NRI and to permit the nation to meet the challenges to the national (and indeed global) food, fiber, and natural-resources system.

THE NRI'S MISSION

A successful grants program contains elements of value, relevance, quality, fairness, and flexibility. The committee found that the proposals to the NRI and the research conducted by scientists who receive NRI grants are of high quality. That finding is based on the results of the committee's survey of applicants, awardees, administrators of land grant institutions, and industry; the views of former chief scientists and individuals from federal agencies; and the personal perspectives of committee members and their colleagues. Through conscientious stewardship, the NRI has been successful in generating fundamental and applied research and fostering the development of future scientists with strong backgrounds in food, fiber, and natural resources.

The committee recommends that a major emphasis of the NRI continue to be the support of high-risk research with potential long-term payoffs. Much of this research would be classified as fundamental in the traditional use of this term. The NRI also should continue to emphasize the importance of multidisciplinary research.

The NRI program is credited with important contributions to fundamental and applied research. The distinction between fundamental (or basic) and applied research often is unclear in the food, fiber, and natural-resources sector, however. Instead of classifying research arbitrarily as fundamental or applied, it should be thought of as on a continuum with short-, medium-, and long-term objectives identified in any research area. The committee believes that a major emphasis of the NRI should continue to be the support of high-risk research with potential long-term payoffs—the type of research that is unlikely to be funded through other research programs in USDA, other federal agencies, or the private sector. The committee also encourages the NRI to continue to emphasize multidisciplinary research because the problems in the food, fiber, and natural-resources system demand multidisciplinary approaches and collaboration.

The committee recommends that the NRI continue to emphasize its mission of training and education.

The training and education of graduate students and postdoctoral researchers attributable to the NRI program have been valuable. Although grants have been small and of short duration, training appears to have been a major use of NRI funds among university researchers. Strengthening grants provided by the NRI program have had a major impact on the careers and productivity of faculty who otherwise would not receive federal grant support. Furthermore, NRI staff have been successful, particularly in view of the organization's limited resources, in organizing several vehicles to promote public understanding of research in food, fiber, and natural resources.

RESEARCH ACCOUNTABILITY

The committee recommends continuing the process of merit-based peer review as the most effective method of competitively distributing funds for research in food, fiber, and natural resources.

The committee views the NRI as a model of merit-based peer-reviewed research in USDA. Because it uses a competitive review process to rank proposals, however, the NRI remains outside the mainstream USDA culture of formula funding. The successful operation of the peer-review system in the NRI accounts for the high quality of the projects funded. Merit-based peer review has been adopted as the principal criterion of funding of extramural research throughout the federal government and increasingly in universities. It is a consistent, expertise-driven method for allocating research funds fairly and appropriately. Information gathered by the committee indicates that stakeholders in the food, fiber, and natural-resources system hold the NRI peer-review process in high esteem. Some survey respondents indicated that the NRI merit-based peer-review process was as fair as and perhaps more responsive than the review process of other federal research agencies.

The committee recommends that a more effective performance-tracking system be established to improve research accountability.

The committee believes that the NRI could improve its record by documenting the value of research that it funds. The NRI does not keep a definitive record of patents and publications resulting from NRI research. Nor is there a running evaluation of originality and significance of current applications and renewals. Although the committee has found based on its surveys that funded applications are of high quality, the NRI lacks a tracking system of critical factors needed for self-evaluation or effective reporting of research accomplishments to outside groups, which would create a feedback system to establish value.

Every federal research agency faces important challenges in measuring outcomes of research projects, and the NRI is no exception. The committee concluded that the quality of research supported by the NRI is high, but it was unable to scrutinize individual projects extensively because of the absence of a tracking system tailored to tying projects to outcomes. A standardized tracking system needs to be implemented for the NRI program. Such a system would be beneficial both for tracking outcomes and for making the NRI's programs more transparent to stakeholders. The National Research Council has recently released a report, *Evaluating Federal Research Programs,* on accounting for federal outcomes as part of the Government Performance and Results Act mandate. The NRI should use the recommendations in that report.

The committee recommends implementation of an internal information system that generates data on current operations of the NRI.

The committee found it difficult to follow year-to-year changes in funding areas and to generate numbers to measure effort by project and category outcome. The committee's requests for information generated more work by the NRI professional staff than should have been required. The committee believes that those problems were due to deficiencies in the underlying information system itself.

The committee recommends that the NRI Web site be more readily accessible to allow the location of research projects and results with the use of issue-oriented key words and technical terms that are accessible and understandable to all stakeholders.

A number of recommendations reflect directly on the NRI's ability to reach both traditional and new stakeholders. But the needs for transparency, access to the current research agenda, and documentation of past outcomes suggest a substantial expansion in communication strategy. A Web site could be linked to nontechnical summaries, technical abstracts, impact statements, and publications, and to a catalog of current and past funded projects. Such data and communication could be maintained for 10 years to build a timely, comprehensive, and searchable record of research impacts generated by NRI funding.

PRIORITY-SETTING AND ORGANIZATION

The committee has concluded that the priority-setting process of the NRI needs substantial revision. The committee found that parts of the process used by the NRI staff seem unstructured, appear to be unevenly administered across NRI divisions, and are not explicitly linked to the goals and other strategic planning elements of the Research, Education, and Economics Mission Area. Changes in program areas and priorities appear to have occurred primarily in response to the urging of vocal stakeholders rather than as the result of a deliberative priority-setting process. Mechanisms are not well established to evaluate the effectiveness of NRI-funded research as time passes and progress occurs or to delineate how key research outcomes correlate with guiding research goals. The priorities of the NRI do not appear to be linked closely with the priorities of the Agricultural Research Service (ARS) and the Economic

Research Service (ERS), perhaps because the potential cross-functional nature of present research programs is not fully appreciated in either the ARS or NRI administration.

The committee believes that an improved priority-setting process should involve independent input from scientists and informed members of the public. The priority-setting process also should allocate more of the NRI's funds by issue, not by research category. The committee believes that changes in the NRI's organization need to be made. Most important, USDA needs to find a way to enhance the position of extramural research in USDA and to encourage NRI priority-setting to reflect national priorities more clearly.

The committee offers the following recommendations to improve the priority-setting in and the overall effectiveness of the NRI. Other solutions are possible; ultimately, it will be up to USDA, and possibly Congress, to decide how best to address these problems.

The committee recommends that six standing scientific-research review committees be assembled to identify critical issues in each research area. The committee further recommends that the current 26 programs be eliminated and replaced with an issue-based agenda across the six purviews of the committees.

Some NRI divisions have been relatively stable programmatically since their inception, whereas others have seen many program starts and stops. The subdivision of the NRI's six main research areas into 26 programs solely by research "category", in the absence of an overall strategic plan, might have been partly responsible for a lack of critical mass among the NRI's natural stakeholders, particularly because the recommended increases in research funding to $500 million did not materialize.

Several short-term changes in program direction (over 4- to 6-year time frames) have occurred in research areas that would otherwise need about 8–10 years to have an impact. The stop-start nature of some NRI funding commitments over its short history indicates that the NRI has been unable to sustain funding support for some high-risk areas with long-term payoffs—the types of research for which the NRI is ideally suited. The lack of a clear perception of the logic of annual requests for proposals across all 26 programs could be partly responsible for the NRI's inability to attract increased research budgets for its programs. A more logical priority-setting process that relates the NRI's research programs to USDA goals and emerging issues in the food, fiber, and natural-resources system might be effective in demonstrating more clearly the importance of NRI-supported research and lead to increased research budgets.

The sporadic development of the NRI's 26 programs reflects neither a coherent long-term research agenda nor the generation of clear and observable outcomes. The issue-oriented deliberations of the research review committees would form the basis of many annual requests for proposals. The review committees should include scientists from the entire food, fiber, and natural-resources system. A shift in priority-setting might cause a major change in the types of research supported by the NRI.

The committee recommends that the research review committees give special consideration to important problems perceived by the public at large—such as alternative energy, healthfulness of food, food safety, and nutrition (issues at the consumer end of the food system), in addition to the more traditional emphases on productivity, rural economies, and environmental protection.

The likely outcome would be a better distribution of research funds across the entire food, fiber, and natural-resources system and a research agenda more closely aligned with public concerns. The NRI research agenda would thus become more forward-looking and issue-driven.

The committee recommends that a cooperative formal goal and strategy process be instituted in the context of the NRI's role in federal food, fiber, and natural-resources research programs.

The NRI generally complements other USDA activities and does not duplicate other federal research efforts. The NRI actively participates in cross-agency funding opportunities to ensure complementarity of research efforts, but it clearly follows rather than leads in such efforts. Apart from memoranda of understanding and interagency coordination provided by the National Science and Technology Council, no process exists for establishing formal relationships with other federal agencies or for consulting and use stakeholder groups. The committee believes that being smaller than other agencies limits the funding that the NRI can contribute to such cross-agency initiatives.

The National Institutes of Health (NIH), the National Science Foundation (NSF), the Department of Energy (DOE) and the NRI form the backbone of the nation's merit-based peer-reviewed research effort in food, fiber, and natural resources. The NRI is the nation's only merit-based peer-reviewed research program that focuses explicitly on challenges to its system of food, fiber, and natural resources. A comprehensive strategy that required coordination among congressional committees—particularly those with jurisdiction over USDA, NSF, and NIH programs and budgets—would allow an expanded NRI food, fiber, and natural-resources agenda to be coordinated with complementary work funded by NIH and NSF.

The committee recommends that the NRI and other competitive USDA research programs be moved to a new Extramural Competitive Research Service (ECRS) that would report to the under secretary for research, education, and economics (figure 7-1).

The location of the NRI as one component of the Competitive Research Grants and Awards Management Division, rather than on an organizational level equivalent to USDA's two main research agencies (ARS and ERS), suggests that USDA and Congress place a higher priority on formula

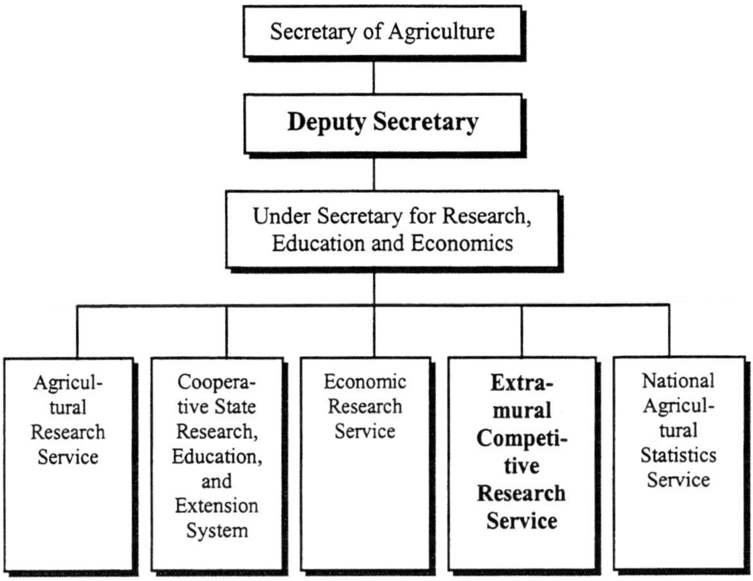

FIGURE 7-1 Recommended Organization of USDA Research, Education, and Economics Mission Area

funds, special grants, and intramural research than on extramural, merit-based peer-reviewed competitive research. The committee believes strongly that unless extramural competitive research is given the same stature organizationally as formula-funded and intramural research in USDA, it will remain difficult for the NRI program to achieve its mission.

The committee believes that the NRI has suffered as a program in an agency—the Cooperative State Research, Education, and Extension Service (CSREES)—that is also responsible for defending and allocating formula funds and special grants. Intramural research is represented by ARS and ERS, which report directly to the under secretary for research, education, and economics, as does CSREES. The committee strongly recommends that extramural competitive research, to achieve critical mass, be given an organizational stature that would allow it to compete effectively for resources with formula funds and special grants and to participate directly in USDA's high-level priority-setting process.

The committee recommends the establishment of an Extramural Advisory Board (12–14 members) that represents NRI stakeholders and has a non-USDA chair.

Funding has been unevenly allocated among the NRI's divisions since its initiation. No substantial changes in the proportions of funding allocated to the divisions have occurred, even though the nature of food, fiber, and natural resources has changed since 1991. Funding allocations do not appear to have distinguished between traditional and emerging areas in food, fiber, and natural resources.

The current NRI Board of Directors provides necessary administrative oversight of the NRI program and can be used to link the NRI with USDA's other research organizations. The Board of Directors is not responsible for providing guidance on scientific or technologic priorities, providing a forum for stakeholder concerns, or measuring research outcomes and evaluating NRI operations. An external advisory board of some type is necessary to handle those responsibilities.

The Advisory Board would advise and assist the chief scientist in identifying fundamental issues and future strategies to meet the greatest needs. It would represent scientists and engineers, deans of land grant and non-land grant institutions, industry across the entire food and fiber system, commodity and farm groups, consumer groups, and 1890 colleges. Ex officio members would include select program managers at NIH and NSF and the NRI chief scientist. Board members would serve 3-year terms on a staggered, rotating basis with a maximum of two terms. The board would be appointed by the secretary of agriculture.

In the committee's opinion, an external Advisory Board is critical to the successful functioning of the NRI. Stakeholder contact, the advocacy of extramural research inside and outside USDA, measurement of research outcomes, and continuing evaluation of NRI operations (including the peer-reviewed project-selection system) would ensure thoroughness, objectivity, and transparency. A visible, mandated external Advisory Board would bring renewed energy and focus to an expanded NRI effort and would provide Congress with an objective appraisal of NRI efforts.

The committee recommends that the position of chief scientist be a full-time, permanent 5-year position, with the option of one 5-year renewal, chosen by the secretary of agriculture with the consultation, recommendations, and advice of the newly created Extramural Advisory Board. The chief scientist would be the administrator of ECRS.

The current responsibilities of the NRI chief scientist are equivalent to a full-time position. A part-time revolving chief scientist cannot meet the strategic-planning, priority-setting, and communication needs of an effective NRI. Although past chief scientists have done excellent work, having a part-time chief scientist impedes continuity in accountability and leadership and counters successful long-range planning and followup and consistent stakeholder involvement.

The necessary duties of the chief scientist-administrator of ECRS, in addition to those now assigned within the NRI, would include directing the program and developing a definitive strategic plan and advocacy for the NRI program. The chief scientist could also take the lead in changing the culture of the NRI from a program-based to an issue-based research agenda. The full-time chief scientist would report directly to the under secretary and would play a major role in setting the nation's federal food, fiber, and natural-resources research agenda.

The committee recommends that each of the six mandated areas of research emphasis be led by a half-time associate chief scientist with a 2-year rotation. Each associate would be a scientist from a visible and productive outside research program (figure 7-2).

In recent years, the NRI staff has been stretched to cover its responsibilities, and this has increased the burdens of communication and timeliness on NRI staff at all levels and on the scientists who serve as ad hoc reviewers and panel members. The proposed rotation would allow the chief scientist to recruit a flow of intellectual capital and would provide a mechanism for obtaining input from the population of researchers served by the NRI. The full-time chief scientist plus the six associate chief scientists would have the time and resources to carry out long-term analyses of research needs in the context of issues rather than programs, as is now the case. This recommendation highlights the importance of establishing and maintaining a scientifically based research agenda. The associate chief scientists would complement the division directors, program managers, and volunteer panel leaders.

A number of factors could account for the fact that USDA's research agenda has struggled over the last decade. The committee understands current budget constraints and understands that the implementation of some of its recommendations would increase personnel and operating costs. We believe strongly, however, that substantial changes are needed to ensure the

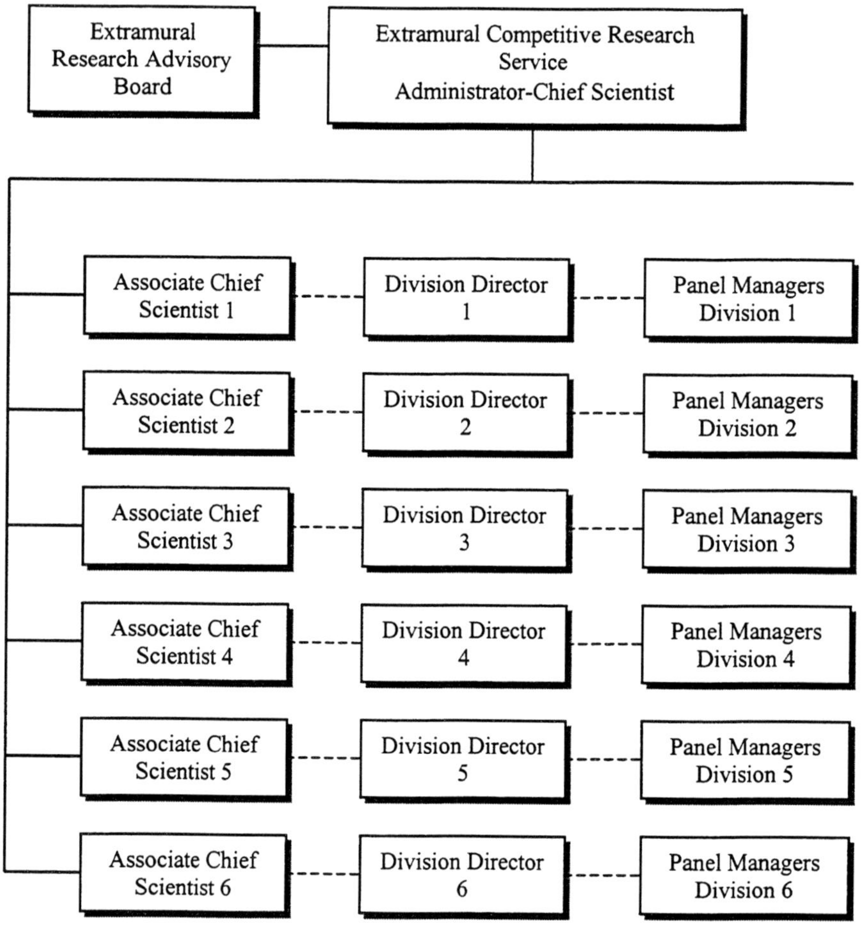

FIGURE 7-2 Recommended Organization of USDA Extramural Competitive Research Service (New NRI)

future success of merit-based peer-reviewed research in food, fiber, and natural resources.

FUNDING

The committee recommends that grant awards be immediately increased to an average of $100,000 per year (total costs) over 3 years.

NRI research grants are much smaller and shorter than grants supporting similar types of research at NSF, NIH, and DOE. Continued underfunding of NRI research grants relative to those of other federal research agencies will tend to discourage new researchers outside the traditional food and fiber system from applying for NRI grants—one original goal of the NRI. It might also cause highly qualified scientists who have received NRI support to apply for research funds from other sources and even redirect their research away from issues important to the food and fiber system. That could lead to a decrease in the overall quality of food, fiber, and natural-resources research.

The proposed increase would solidify the stakeholder foundation of the NRI and prepare it to receive additional funds. The committee recognizes that without an increase in the NRI's total budget (as recommended strongly by this committee), the increase in size and duration of grants would reduce the number of grants and perhaps cause hardship among investigators who have depended on NRI funding to sustain their research programs. However, continued underfunding of individual research grants would reduce the aggregate impact of the NRI's competitive funding. The number of current proposals is lower than in the past, and stakeholder support appears to be waning. An increase in the size and duration of grants would enable the scientific community to attack issues in food, fiber, and natural resources by preparing proposals that require multi-investigator and multidisciplinary teams of researchers. Increased size and duration of grants would allow researchers to carry out projects as planned without narrowing their scope to fit a shorter period and smaller amount. Finally, increased size and duration of grants would attract new, creative proposals from researchers who are now outside the traditional food and fiber system. The latter was one of the key reasons for instituting the NRI, and it continues to be a worthwhile objective. To achieve it, the NRI must provide realistic funding levels to continue to attract the best and the brightest students and investigators to food, fiber, and natural-resources research.

The NRI should benchmark the funding level and duration of its grants to those of the other federal merit-based peer-review agencies that support research. NSF and NIH support competitive research projects in some of the same basic science and engineering areas as the NRI, that complement food, fiber, and natural-resources research. The challenge is to keep the best intellectual capital engaged in the NRI's scope of issues.

The committee recommends that the NRI's overhead limit be immediately replaced with indirect-cost standards that are used by other federal research agencies.

When it established the NRI program in 1991 Congress imposed a 14% limit on the amount of indirect costs that can be charged as a percentage of the total award.[1] The 14% limit was replaced by a 19% limit[2] in FY 2000 as part of the Agricultural Research, Extension, and Education Reform Act of 1998. Although the increase from 14% to 19% reduces the gap between overhead rates on NRI grants and rates on grants awarded by other federal agencies, overhead rates for most academic and private-sector research institutions are significantly higher than the 19% limit currently allowed. Average overhead rates for NSF's Biology Directorate, for example, are approximately 45% of the modified total direct costs of the award—nearly double the NRI limit. The committee is not aware of any other federal merit-based peer-reviewed research program with such a congressionally mandated limit on overhead rates.

Presumably, the motivation for setting such a limit was to increase the percentage of NRI research funds spent on research activities. However, such a mandated cap on overhead may have a negative effect on the NRI program because it causes some institutions (especially those from outside the traditional applicant community) to discourage their researchers from submitting proposals to the program. Because the committee did not address this issue in its survey, it was not able to estimate the magnitude of this effect on the NRI program. However, the committee is aware of one research institution that prohibits its scientists from submitting proposals to the NRI because the low overhead rates do not cover the true institutional costs associated with such research and because its auditors require consistency among all incoming grants. Other institutions discourage their researchers from submitting proposals by requiring that the researchers (or their departments) use other funds to make up the difference between mandated low overhead rates and the established rates used by other federal agencies. This is especially problematic for smaller institutions where researchers do not have the flexibility to balance low-overhead grants against other sources of unrestricted funds. These factors also may have a disproportionate impact on institutions (or departments) from outside the traditional food, fiber, and natural-resources system because they do not have a historic association with the USDA, and hence, may be less willing to accept a low overhead rate that is unique to USDA-sponsored research.

The committee believes that Congress could help broaden the scope of NRI researchers beyond the traditional food, fiber, and natural-resources system—one of the original goals of the program—by allowing the NRI to use the same negotiated overhead rates used by other federal agencies. This action, together with the increased grant amounts recommended previously, would make the NRI a more attractive source of funding to all institutions and researchers, and hence could encourage proposals from researchers from outside the traditional food, fiber, and natural-resources system.

---

[1]This limitation is equivalent to 0.16279 of the total direct costs of an award.

[2]This limitation is equivalent to 0.23456 of the total direct costs of an award.

The committee recommends that by 2005 the NRI budget be increased to a level equivalent (adjusted for inflation) to the $550 million recommended by the NRC in 1989—but only if recommended changes in priority-setting, documentation, and organization are put into place.

Inadequate funding of the NRI has significantly limited its potential and placed the program at risk. A substantial increase in funding will ensure a robust public research effort that can significantly transform the nation's food, fiber, and natural resources system in response to critical needs in agricultural productivity, environmental health, and societal well-being.

In its 1989 report *Investing in Agricultural Research,* the NRC called for expanding competitive research within the USDA and establishing the NRI, with a proposed funding increase to $550 million within one year, if possible. Congress responded in 1990 by authorizing $500 million for the NRI by 1995. The committee strongly re-affirms the previous NRC recommendation. Considering inflation alone, $550 million in 1989 is equivalent to approximately $700 million in current (2000) dollars. Assuming conservatively that future annual rates of inflation rate will be roughly 3%, the equivalent size of the NRI budget would be approximately $800 million in 2005. The committee believes that attaining this level would be an important step in re-energizing the national food, fiber, and natural resources research complex—which in turn, would result in major benefits to the nation. After reaching this budget level, the future growth of the NRI budget should be evaluated and compared with the growth in the budgets of complementary research programs in NSF, NIH, and DOE, as suggested by the committee's earlier recommendation to benchmark the amount and length of NRI grants against such research programs.

To illustrate the potential impact of such a budget increase, the committee has done some rough calculations to estimate the number of NRI research awards that could be made with such a budget. If 10% of the budget is spent on "strengthening grants"[3] and administrative costs are 4% of the budget,[4] approximately $700 million would be available for competitive research grants in 2005. Assuming that NRI average award amounts are benchmarked against awards made by other federal programs such as NSF's Biology Directorate and DOE's Energy Biological Sciences (as recommended previously) and that these average awards amounts increase at roughly 3% per year from the current annualized amount of $100,000, the average 3-year NRI grant would be approximately $350,000 in 2005. This would correspond to approximately 2,000 grants to be awarded each year by 2005 (with a total of 6,000 grants being supported at any one time because the grants would be for three years). If the NRI were to adopt an "issue-based" research agenda (as recommended previously) that includes roughly the same number of issues as were identified

---

[3]Congress specified in the Food, Agriculture, Cosnervation, and Trade Act (FACTA) that research and education strengthening grants be at least 10% of NRI's budget.

[4]Congress specified in FACTA that NRI administrative costs be less than 4% of NRI's budget.

by the committee in its list of emerging research issues—about 50 issues—this would correspond to about 40 grants awarded each year for each issue (including new submissions, renewals, and re-submissions). The committee believes strongly that an effective issue-based research program in the food, fiber, and natural resources area requires this level of investment.

The committee recognizes that this recommendation would require a major increase in funding for the NRI. To put the recommendation in context, however, it is useful to compare the estimates given above with funding levels for other research programs within USDA and for other federal agencies. For example, in FY 1999 the USDA Agricultural Research Service's budget was nearly $800 million and USDA formula funds totaled $541 million. NSF's and NIH's budgets for FY 2000 are $3.9 billion and $17.9 billion, respectively, and the budget for DOE's Office of Science for FY 2000 is $2.8 billion, according to a 1999 article in Science. Given these data, the committee does not think it unreasonable to expect that a competitive research program explicitly focused on high-priority issues in food, fiber, and natural resources—essential elements to future national security and stability—be funded at approximately $800 million by 2005. As stated previously, this figure is essentially a re-affirmation of the NRC's 1989 recommendation to increase competitive research funding at USDA to $550 million.

The committee believes that the recommended increase in funding should take place incrementally as the various changes recommended earlier in this report are put into place. The ability to utilize large amounts of new funding effectively will be compromised unless recommended changes to the priority-setting process and NRI's organization are implemented.

SUMMARY

The committee found the NRI's current peer-reviewed research to be of high quality and value but believes that much could be done to characterize the quality and value more concretely and to communicate that information to the stakeholders in the NRI better. The committee found the NRI priority-setting process to be lacking. Specific structural changes were recommended to remedy that deficiency.

The committee found that the NRI's research agenda complements other USDA activities and those of other federal agencies, the states, and the private sector. However, the current size, structure, and diffuse agenda make effective complementarity difficult. The committee recommends changes in process and priority-setting to help buttress this NRI responsibility.

Finally, the committee set forth comprehensive organizational and funding changes so that the NRC's vision for food, fiber, and natural-resources research could be achieved. A combination of restructuring and substantially increased funding could provide USDA and the nation with the critical fundamental merit-based peer-reviewed research base that will be required to meet the food, fiber, and natural-resources challenges of the 21st century.

Appendixes

Appendix A

Section 1615 of the Food, Agriculture, Conservation and Trade Act of 1990

SEC. 1615. NATIONAL COMPETITIVE RESEARCH INITIATIVE.

(a) INITIATIVE ESTABLISHED- Subsection (b) of section 2 of Public Law 89-106 (7 U.S.C. 450i) is amended--

(1) by inserting 'COMPETITIVE GRANTS- (1)' after '(b)'; and (2) by striking the third sentence and all that follows and inserting the following new paragraphs:

'(2) HIGH PRIORITY RESEARCH- For purposes of this subsection, the term 'high priority research' means basic and applied research that focuses on both national and regional research needs (and methods to transfer such research to onfarm or inmarket practice) in--

'(A) plant systems, including plant genome structure and function; molecular and cellular genetics and plant biotechnology; plant-pest interactions and biocontrol systems; crop plant response to environmental stresses; unproved nutrient qualities of plant products; and new food and industrial uses of plant products;

'(B) animal systems, including aquaculture, cellular and molecular basis of animal reproduction, growth, disease, and health; identification of genes responsible for improved production traits and resistance to disease; improved nutritional performance of animals; and improved nutrient qualities of animal products, and uses, and the development of new and improved animal husbandry and production systems that take into account production efficiency and animal well-being, and animal systems applicable to aquaculture;

'(C) nutrition, food quality, and health, including microbial contaminants and pesticides residues related to human health; links between diet and health; bioavailability of nutrients; postharvest physiology and practices; and improved processing technologies;

'(D) natural resources and the environment, including fundamental structures and functions of ecosystems; biological and physical bases of sustainable production systems; minimizing soil and water losses and sustaining surface water and ground water quality; global climate effects on agriculture; forestry; and biological diversity;

'(E) engineering, products, and processes, including new uses and new products from traditional and non-traditional crops, animals, byproducts, and natural resources; robotics, energy efficiency, computing, and expert systems; new hazard and risk assessment and mitigation measures; and water quality and management; and

'(F) markets, trade, and policy, including optional strategies for entering and being competitive in overseas markets; new decision tools for onfarm and inmarket systems; choices and applications of technology; technology assessment; and new approaches to rural economic development.

'(3) TYPES OF GRANTS- In addition to making research grants under paragraph (1), the Secretary may conduct a program to improve research capabilities in the agricultural, food, and environmental sciences and award the following categories of competitive grants:

'(A) Grants may be awarded to a single investigator or coinvestigators within the same discipline.

'(B) Grants may be awarded to teams of researchers from different areas of agricultural research and scientific disciplines.

'(C) Grants may be awarded to multidisciplinary teams that are proposing research on long-term applied research problems, with technology transfer a major component of all such grant proposals.
'(D) Grants may be awarded to an institution to allow for the improvement of the research, development, technology transfer, and education capacity of the institution through the acquisition of special research equipment and the improvement of agricultural education and teaching. The Secretary shall use not less than 25 percent, and not more than 40 percent, of the funds made available for grants under this subparagraph to provide fellowships to outstanding pre-and post-doctoral students for research in the agricultural sciences.

'(E) Grants may be awarded to single investigators or coinvestigators who are beginning their research careers and do not have an extensive research publication record. To be eligible for a grant under this subparagraph, an individual shall have less than 5 years of post-graduate research experience.

'(F) Grants may be awarded to ensure that the faculty of small and mid-sized institutions who have not previously been successful in obtaining competitive grants under this subsection receive a portion of the grants.

'(4) TERM- The term of a competitive grant made under this subsection may not exceed 5 years.

'(5) DIRECTOR- The Secretary shall appoint a director for the grant program authorized by this subsection. The Secretary, acting through the director, shall be responsible for the overall direction of the grant program and implementation of general policies respecting the management and operation of programs and activities in the program.

'(6) PARTICIPATION IN GRANT PROCESS- In seeking proposals for grants under this subsection and in performing peer review evaluations of such proposals, the Secretary shall seek the widest participation of qualified scientists in the Federal Government, colleges and universities, State agricultural experiment stations, and the private sector.

'(7) CONSTRUCTION PROHIBITED- A grant made under paragraph (1) may not be used for any purpose for which a grant may be made under subsection (d) or for the planning, repair, rehabilitation, acquisition, or construction of a building or facility.

'(8) MATCHING FUNDS-

'(A) Except as provided in subparagraph (B), the Secretary may not take the offer or availability of matching funds into consideration in making a grant under this subsection.

'(B) In the case of grants under paragraph (3)(D), the amount provided under this subsection may not exceed 50 percent of the cost the special research equipment or other equipment acquired.

'(9) ANNUAL REPORT- The Secretary shall transmit to Congress an annual report describing the policies, priorities, and operations of the grant program authorized by this subsection during the preceding fiscal year. The report shall--

'(A) include a description of the progress being made to comply with subsection (j); and

'(B) be transmitted not later than January 1 of each year.

'(10) AUTHORIZATION OF APPROPRIATIONS- There are authorized to be appropriated to carry out this subsection $150,000,000 for fiscal year 1991, $275,000,000 for fiscal year 1992, $350,000,000 for fiscal year 1993, and $400,000,000 for fiscal year 1994, and $500,000,000 for fiscal year 1995, of which each fiscal year--

'(A) not less than 10 percent for fiscal year 1991, 20 percent for fiscal year 1992, and 30 percent for fiscal year 1993 and each fiscal year thereafter shall be available to make grants for research to be conducted by multidisciplinary teams;

'(B) not less than 20 percent shall be available to make grants for research to be conducted by persons conducting mission-linked systems research;

'(C) not less than 10 percent shall be available to make grants under subparagraphs (D) and (F) of paragraph (3) for awarding grants in research and education strengthening and research opportunity;

'(D) not more than two percent may be used for equipment grants under subparagraph (3)(D); and

'(E) not more than four percent may be retained by the Secretary to pay administrative costs incurred by the Secretary in carrying out this subsection.'.

(b) ADMINISTRATIVE PROVISIONS- Such section is further amended by adding at the end the following new subsections:

'(j) EMPHASIS ON SUSTAINABLE AGRICULTURE- The Secretary of Agriculture shall ensure that grants made under subsections (b) and (c) are, where appropriate, consistent with the development of systems of sustainable agriculture. For purposes of this section, the term 'sustainable agriculture' has the meaning given that term in section 1404(17) of the National Agricultural Research, Extension, and Teaching Policy Act of 1977 (7 U.S.C. 3103(17)).

'(k) REPORTS- The Secretary of Agriculture shall prepare and submit to Congress on January 1 of each year a report on awards made under subsections (b) and (c) during the previous fiscal year.

'(l) CONSULTATION WITH TECHNOLOGY BOARD- The Secretary of Agriculture may consult with the Agricultural Science and Technology Review Board regarding the policies, priorities, and operation of subsections (b) and (c).'.

(c) STYLISTIC AMENDMENTS- Such section is further amended—

(1) by striking 'SEC. 2. (a)' and inserting the following:

'SEC. 2. COMPETITIVE, SPECIAL, AND FACILITIES RESEARCH GRANTS.
'(a) ESTABLISHMENT OF GRANT PROGRAM- ';

(2) in subsection (d), by inserting 'FACILITIES GRANTS- ' after '(d)';

(3) in subsection (e), by inserting 'RECORD KEEPING- ' after '(e)';

(4) in subsection (f), by inserting 'LIMITS ON OVERHEAD COSTS- ' after '(f)';

(5) in subsection (g), by inserting 'AUTHORIZATION OF APPROPRIATIONS- ' after '(g)';

(6) in subsection (h), by inserting 'RULES- ' after '(h)';and (7) in subsection (i), by inserting 'APPLICATION OF OTHER LAWS- ' after '(i)'.

Appendix B

Survey on the US Department of Agriculture's National Research Initiative (NRI) Competitive Grants Program

1. Are you familiar with the NRI program?

a) Yes, very familiar _______
b) Somewhat familiar _______
c) Not familiar _______ (Please answer Question 8 and any others you consider appropriate.)

Comments:

2. In your view, has the program contributed to the mission of generating fundamental and applied research, and future scientists for agriculture?

a) Yes ______ for the most part.
b) Yes ______ but less than I expected.
c) No ______ its promise remains largely unfulfilled.

Comments; Examples:

3. Is the NRI peer review process fair?

a) Yes_______
b) No _______

Comments:

4. As a NRI grant recipient/applicant, how did this affect your career? (If this question does not apply to you, but you know someone whose career is or has been affected, please fill in the following.)

	Greatly	Some	Very Little
a) Promotion impact	_____	_____	_____
b) Tenure impact	_____	_____	_____
c) Publication(s)	_____	_____	_____
d) Patent(s)	_____	_____	_____
e) Career development of undergrad. students	_____	_____	_____
f) Career development of your graduate students	_____	_____	_____
g) Career development of your post-doctoral fellows	_____	_____	_____
h) Ability to pursue independent (curiosity-driven) research	_____	_____	_____

Other (comments):

5. Has the NRI enabled you or your institution/company to obtain other funds (leveraging)?

a) Yes_____
b) No _____

Comments; Examples:

6. Has the NRI contributed to development of human resources in food and agriculture?

a) Yes_____
b) No _____

Comments; Examples

7. Has the NRI program resulted in any major benefits to the U.S. food and agriculture system?

a) Yes_____
b) No _____

Comments; Examples:

8. Are the areas of congressionally mandated funding appropriate? These areas are: plant systems; animal systems; nutrition, food quality and health; natural resources and the environment; processes for adding value and new products; and markets, trade and policy.

a) Yes_____
b) No _____ (Please explain in the comments area)
c) What two areas of research are most important to you, in order of priority (1 = highest)?

Comments:

9. If the NRI received a significant increase in appropriations, how should the additional funds be used? Rank the following (1 = highest; 5 = lowest).

a) Expand into new areas _______
b) Increase size of awards _______
c) Increase duration of awards _______
d) Increase number of submission dates per year _______
e) Other _______

10. Should the NRI continue? Mark all that apply.

	Very Much	Some-what	Not Much
a) Is the NRI continuance important to you	_____	_____	_____
b) Is the NRI continuance important to your institution	_____	_____	_____
c) Is the NRI's continuance important for the U.S.?	_____	_____	_____
d) Is the NRI an important part of the USDA research portfolio?	_____	_____	_____
e) Is the overhead rate (now 19%) *acceptable*? *a deterrent*? (please circle one and then respond)	_____	_____	_____

11. How would you improve the NRI? Mark all that apply.

a) Change program areas	Yes_____	No_____
b) Change application process	Yes_____	No_____
c) Change review process	Yes_____	No_____

If "yes", please elaborate and give examples:

Appendix C

External Views of the NRI

The committee sought to gather impressions and systematic data on the functioning of the NRI from knowledgeable constituent groups through three mechanisms: a survey, interviews with former NRI chief scientists, and formal testimony of NRI stakeholders. As a first comprehensive effort to assess the functioning of the NRI, surveys (see appendix B) were mailed to deans and directors of land grant and non-land grant institutions, recipients of NRI grants, nonrecipients in the applicant pool, and representatives of industry. In October 1998, the committee conducted interviews with four former chief scientists. The committee also received testimony from a wide range of stakeholders, including professional societies, nonprofit research institutes, industry, universities, experiment stations, farm organizations, and federal agencies. This appendix summarizes the results of all those efforts.

SURVEY RESULTS

Four specific groups were identified by the committee for its survey of the NRI: recipients of NRI grants, nonrecipients of NRI grants, administrators of land grant institutions, and industry. The NRI provided lists of recipients and nonrecipients of grants for 1995-1997. Names and addresses of administrators of land grant and non-land grant institutions were supplied by the National

Association of State Universities and Land Grant Colleges. The committee determined the remaining contacts, including representatives of medium to large companies. The survey was sent to every tenth name on awardee and nonrecipient (declined at least twice) lists.

Questions formulated by the committee were mailed by the NRI to awardees and nonrecipients in August 1998 (see appendix B). The National Research Council staff mailed the survey to the remaining groups. Those surveyed were asked to select from among the given responses and were also encouraged to provide additional written comments. Replies were collated by the staff, and the committee reviewed the responses, evaluated the results, and summarized recommendations from those surveyed. Although the survey is not statistically representative of the NRI applicant and awardee populations, the large number of respondents is indicative of the views of those with experience with the NRI.

Response Rates and General Themes

Table A2–1 presents the response rates of all surveyed groups that were tracked in the survey (for example, researchers at federal laboratories were included among NRI awardees and nonrecipients, not tracked as a separate group). The survey had a response rate of at least 50 % for three of the four groups, industry was the exception. Table A2-2 breaks down the response rates for recipients and nonrecipients by NRI program area.

TABLE A2–1 Response Rate by Group Surveyed

Group Surveyed	No. Sent	No. Received	Response Rate, %
Awardees	203	141	69.5
Nonrecipients	102	51	50.0
Land Grant Institutions	85	60	70.6
Industry	142	37	26.1
Total	532	289	54.3

TABLE A2-2 Responses of Awardees and Nonrecipients by NRI Research Area

Research area	No. Awardees		No. No-recipients	
	Sent	Received	Sent	Received
Natural Resources and the Environment	24	20	21	7
Nutrition, Food Safety, and Health	13	10	9	5
Animals	36	25	24	11
Plants	46	29	12	6
Markets, Trade, and Rural Development	9	6	5	3
Enhancing Value and Use of Agricultural and Forest Products	16	9	7	4
Pest Biology, Biological Control, and Integrated Pest Management	35	25	19	12
Agricultural Systems Research	3	2	3	2
Strengthening Programs	20	14	2	1
NSF/DOE/NASA/USDA Joint Program on Terrestrial Ecology and Global Change (TECO)	1	1	—	—
Total	203	141	102	51

Nearly all respondents indicated that the NRI program had contributed to generating fundamental and applied research, and training future scientists for agriculture (see question 2). A large majority of respondents indicated that the NRI had contributed to the development of human resources in food and agriculture, specifically in career development and predoctoral and postdoctoral training (see question 6). In addition, an overwhelming percentage of those surveyed in all four target groups believed that the NRI program had resulted in major benefits to the US food and agriculture system (see question 7). Virtually all respondents viewed continuation of the NRI program as essential (see question 10).

An overwhelming percentage of NRI recipients and a majority of nonrecipients indicated that the NRI uses a fair peer-review process to select proposals for funding (see question 3). Respondents who had review-panel experience in other competitive grants programs were especially complimentary of the NRI process. Some nonrecipients and administrators of land grant institutions criticized some elements of the peer-review process, including panel composition, the single yearly application, the long response times, and the length of panel terms.

Most of respondents believed that the congressionally mandated program areas are appropriate (see question 8). A few respondents that the areas overlapped or did not represent their research interests. Several respondents favored high-risk, high-reward projects and suggested greater emphasis on interdisciplinary proposals.

Questions 9 and 11 addressed improvements in the NRI program. Question 9 asked how additional money should be used if the NRI received a large increase in appropriations. Question 11 asked respondents how they would improve the NRI. The majority response to question 9 was to increase the size and duration of awards, followed by expansion into new areas. In response to question 11, awardees were generally satisfied with program operations. Their responses centered on refining the review process, they recommended less turnover among the review panels and more rapid and more frequent reviews. Awardees also expressed concern about reviewers competing for grants themselves in their research areas. Nonrecipients expressed concerns about the review process, especially the selection of panel members and reviewers and the need to avoid an "old boy network". They also favored more rapid and more frequent reviews. Several respondents suggested a preproposal process to shorten evaluation time.

Many respondents expressed the view that total funding of the program was insufficient and that awards were too short, too few, and too small (see question 11). The low overhead rate was cited by some respondents but was not a major concern.

The survey was limited in scope and statistical significance. The results reflect the views only of the respondents. Most respondents clearly had a personal interest in the continuation and expansion of the NRI. However, the high response rate, the thoroughness of the responses, and the numerous suggestions for improvement reflect the high importance of the NRI to these members of the research community.

Detailed Summary of Survey Data

The following sections summarize the survey results for the groups tracked in the survey for each of the 11 survey questions. For each question (sometimes each group), a tabular summary of responses is followed by an overview of the written comments provided by some respondents. Except where noted, narrative descriptions refer only to written comments and thus reflect the views of respondents who took the time to provide them; the narrative descriptions do not necessarily reflect the general views of all those who completed the survey.

1. Are you familiar with the USDA/NRI program?

	Awardees	Nonrecipients	Land Grant	Industry
Yes, very familiar	117	44	47	6
Yes, somewhat familiar	22	7	12	15
Not familiar	2	—	1	15

NRI Awardees:

Twenty-three indicated that they had been involved in the NRI activities either as panel members, panel managers, or ad hoc reviewers. Some awardees had been denied NRI awards in the past.

Nonrecipients:

This group included scientists who in 1995 - 1997 had been denied NRI awards at least twice. The nonrecipient included researchers who had received NRI grants in the past. Six applicants indicated that they participated in the NRI process as panel members or reviewers.

Land Grant:

One respondent served as a reviewer, one as a panel manager, and one as a panel member. One indicated that his university was a big source of reviewers for the NRI. Two respondents wrote that researchers from their institutions applied for NRI grants and some were successful in receiving them. One person felt that because of the way funds were allocated and the method of evaluation, program activities were widely dispersed and funded for a short time; therefore, the NRI does not nearly reach the potential one would demand."

Industry:

Only a few respondents made comments about the NRI. Three respondents were not interested in learning about the program, because of the confidentiality of their research or because the application and review process was too slow.

2. *In your view, has the program contributed to the mission of generating fundamental and applied research, and future scientists for agriculture?*

	Awardees	Nonrecipients	Land Grant	Industry
Yes, for the most part	111	30	33	12
Yes, but less than I expected	22	13	24	8
No, its promise remains largely unfulfilled	3	5	—	—

Awardees:

Although some respondents commented that the program focused too much on fundamental or applied research, the general consensus was that despite limited funding, the NRI contributed greatly to fundamental research and to the training of scientists (graduate students and postdoctoral scientists.) In many instances the program provided funds to innovative and valuable research that otherwise would not have been performed. Many respondents considered the NRI crucial to US agriculture. Almost all comments indicated that more funding would make the program more effective.

Nonrecipients:

Only a few researchers made positive comments about the NRI. Some indicated that because of low funding many program areas were not adequately addressed. As a result, many researchers were discouraged from applying for NRI funding. Some felt that only "cutting-edge" and "politically popular" ideas were funded. Others felt that reviewers were not "adventurous" and avoided more innovative projects. Several commented that there was too little emphasis on applied research; one stated the opposite.

Land Grant:

Respondents agreed that NRI funds were too low, which, in combination with the low success rate, discouraged many researchers from participating in the program. Many agreed that the NRI was generating important fundamental research and allowed training of graduate students. However, comments were also made that more funding should be available for applied research. There were only a few comments that more awards should be made to 1890 institutions.

Industry:

A few respondents wrote that the program was underfunded; two felt that the NRI was successful, and one felt that it contributed to advanced knowledge about pork quality and post-mortem physiology.

3. *Is the NRI peer review process fair?*

	Awardees	Nonrecipients	Land Grant	Industry
Yes	125	39	49	16
No	4	14	8	—

Awardees:

A number of respondents indicated that the process was fairer than that of NIH or NSF. Those who served on NRI panels wrote highly about the fairness and honesty of the process, in which the best science is awarded. Many respondents considered reviewers' comments appropriate and constructive. Some felt that reviewers were too conservative in their approaches to new ideas. Some felt that proposals were reviewed by competitors; two had been asked to review proposals while their own proposals were being reviewed within the same sections. Comments were also made that the NRI should ensure continuity on the panels to avoid inconsistency in evaluating grants from year to year. Four respondents marked both "yes" and "no".

Nonrecipients:

Most of the comments were critical of the reviewers and panel members. Some applicants had a perception that panels do not spend enough time to review proposals and rely too heavily on reviewers' comments. Sometimes the reviewer and the applicant compete for the same money. Others felt that there was too much favoritism in selecting panel members and reviewers and that these groups represented the "old boy" network; there is also favoritism in funding of some program areas. Many excellent proposals do not receive funding. On the other hand, those who served on NRI panels attested to the fairness of the process—each project receives a lot of attention and is discussed extensively. One panel member mentioned that projects were discussed in the order in which they were received at the NRI, so the earlier-arriving proposals received better and fresher consideration; this person recommended numbering projects randomly instead of sequentially. Two respondents marked both "yes" and "no."

Land Grant:

A majority of comments showed negative perception of the review process. Respondents felt that panel members were biased against applied research and some categories of applicants and institutions. One person used the term "old boy network". Expertise of reviewers and panel members was questioned as well. Some felt that a broader-based review process should be implemented for the interdisciplinary research and that more unconventional ideas should be funded.

Industry:

Three comments were made in this section: the process is too political and scientists used grant reviews to prepare for the next grant, the process is fair (on the basis of discussions with other researchers), and expectations are higher than reality of needs and ability to deliver.

4. *Has your career been affected by applying for or receiving an NRI grant? (If this question does not apply to you, but you know someone whose career is or has been affected, please fill in the following.)*

Awardees:

	Greatly	Some	Very Little
Promotion impact	65	43	17
Tenure impact	50	30	29
Publication(s)	89	43	4
Patent(s)	7	22	58
Career development of undergrad. students	35	51	29
Career development of your graduate students	84	21	15
Career development of your post-doctoral fellows	71	19	17
Ability to pursue independent (curiosity-driven) research	82	32	16

About 50% of respondents felt that NRI grants contributed to their promotions. About 60%indicated that they were able to pursue independent research that led to publications. Many respondents emphasized the NRI's role in creating research and training programs; for most of them, NRI grants were crucial in establishing their research (often curiosity-driven) programs that attracted many graduate students and postdoctoral scientist (60% of respondents indicated great NRI impact in this area). Many of these grants allowed young researchers to manage and direct research projects and prepared them to run their own laboratories. The NRI is considered essential to fundamental and basic research. Again, many comments were made about too low funding.

Non-recipients:

	Greatly	Some	Very Little
Promotion impact	14	12	16
Tenure impact	12	8	17
Publication(s)	18	11	15
Patent(s)	2	3	24
Career development of undergrad. Students	6	14	19
Career development of your graduate students	19	18	8
Career development of your post-doctoral fellows	17	9	14
Ability to pursue independent (curiosity-driven) research	15	18	14

This group of respondents made very few comments. A couple of researchers expressed the opinion that the NRI is "absolute necessity" and that receiving an NRI award was "one of the most important milestones a faculty member can achieve." Some felt that NRI funds allowed them to conduct fundamental research that wouldn't have been possible otherwise. However, others wrote that applying for NRI grants meant nothing unless funds were received and that not receiving funds had rather adverseeffects on their careers.

Land Grant:

	Greatly	Some	Very Little
Promotion impact	16	14	9
Tenure impact	15	12	10
Publication(s)	20	13	6
Patent(s)	—	9	24
Career development of undergrad. Students	5	17	15
Career development of graduate students	18	16	5
Career development of post-doctoral fellows	13	21	6
Ability to pursue independent (curiosity-driven) research	16	16	8

Very few comments were provided for this section. One respondent felt that NRI funds were not sufficient to support graduate students. Another felt that the NRI was an additional important funding source for the college. One person wrote that the short-term nature of grants did not contribute to an effective research program. Another provided an example of a junior faculty member who went from an NRI postdoctoral position to an assistant professorship; additional funds received from the National Science Foundation (NSF) increased the researcher's chance of promotion and tenure.

Industry:

	Greatly	Some	Very Little
Promotion impact	3	2	4
Tenure impact	2	2	4
Publication(s)	3	2	5
Patent(s)	—	2	7
Career development of undergrad. Students	—	3	6
Career development of graduate students	2	2	5
Career development of post-doc fellows	3	2	5
Ability to pursue independent (curiosity-driven) research	2	2	4

Most of the comments indicated that this question was not applicable. One person used a colleague's career to mark the items.

5. *Has the NRI enabled you or your institution/company to obtain other research funds (leveraging)?*

	Awardees	Nonrecipients	Land Grant	Industry
Yes	78	16	43	2
No	54	30	16	18

Awardees:

Those who received funds were able to continue research supported by NRI grants or to pursue new projects (in some instances, NRI grants supported pilot projects that led to bigger projects supported by industry). Funds were received from state or federal – the US Department of Agriculture (USDA), the National Institute of Health (NIH), NSF, or the Department of Energy (DOE)–industry, commodity and growers groups, and producer organizations. Very expensive commercial computer software was donated to one laboratory. Examples were cited in which NRI grants helped in obtaining funds from Australia and Japan to conduct international collaborative research projects. In other cases, NRI grants helped in obtaining travel funds for an international conference or a fellowship abroad or helped in build a strong program that led to additional faculty and facilities funded by state. Many respondents felt that receiving NRI grants made them more competitive for other funding sources. Only a few comments were made by those who did not receive funds from other sources. Some were in the process of applying for funds and expressed optimism. Some postdoctoral scientists indicated that their NRI grants were not designed to combine with other funds. Only one person stated that the NRI played a minor role for him or her and his or her colleagues in their research programs "and in the end contributes little to them".

Nonrecipients:

Only a few comments were made by this group, mostly by researchers who received funds from other sources. Those sources included state agencies, USDA, growers organizations, and private companies. The respondents felt that receiving NRI grants greatly contributed to their obtaining funds from other sources. Some said that the NRI played a major role in building programs and reputations and that, as the main support of agricultural research, NRI funds were of more value than funds from other sources. One person commented that an established research area did not attract many diverse funding sources.

Land Grant:

Funds were received from various sources: NSF, commodity groups, industry, and as university matches. NRI funding is highly recognized by other agencies and builds a foundation for other funding. It also gives a good start in competing for grants and contracts.

Industry:

Of the two respondents who received funds, only one provided its source: the National Pork Producers Council.

6. *Has the NRI contributed to development of human resources in food and agriculture?*

	Awardees	Nonrecipients	Land Grant	Industry
Yes	105	34	53	16
No	10	5	5	1

Awardees:

A vast majority of respondents agreed that NRI grants allowed them to attract and support graduate students and postdoctoral scientists. Training the future generation of researchers is one of the greatest attributes of the NRI. A couple of respondents looked at the human-resources aspect from a different angle. For example, one researcher working on a viral disease of cattle developed an animal model of a closely related viral disease of humans; findings of the research improved meat and milk production and gave insights into a human disease. A similar example was cited by another person, whose work on animal disease led to animal models applicable to human disease, generating proposals to NIH. Other examples were patented processes for strain development, currently being tested in the field; rural development efforts coordinated with county land-use planning, development of a new fertilizer that will have an impact on foodstuffs production and, work on viral diseases that affect agriculture in Montana.

Some of those who did not mark any answer indicated that they had no opinion. Only a few of respondents who marked "no" made comments. One stated that the NRI program was too small and had very little training potential. Another wrote that because of the nature of his or her research – (development of skeletal muscle), no contribution to human resources was made.

Nonrecipients:

Most of the respondents who made comments agreed that the NRI contributed greatly to training of graduate students and postdoctoral scientists, especially in fundamental research. Only one person commented that short-term grants discouraged training of graduate students. Another person wrote that "simply writing or reviewing NRI proposals develops 'human resources,' namely the PIs."

Land Grant:

Support and training of graduate students and postdoctoral scientists were the most common examples of an NRI contribution. Some respondents felt that the NRI contribution was slight and could be improved by funding more applied or mission-oriented proposals.

Industry:

Only one respondent indicated training of graduate students and postdoctoral scientists.

7. Has the NRI program resulted in any major benefits to the U.S. food and agriculture system?

	Awardees	Nonrecipients	Land Grant	Industry
Yes	101	31	49	15
No	6	5	5	2

Awardees:

There was almost complete agreement that the NRI has greatly contributed to the US food and agriculture system. Many respondents made general statements for example that the NRI played a very important role; that improved human resources led directly to professional effectiveness that benefited food and agricultural system, and that the NRI greatly improved fundamental knowledge. Some stated that it was difficult to quantify these benefits, because results of fundamental research did not have immediate applicability. Others cited benefits coming directly from their research projects such as new plant species that require smaller use of pesticides, animal vaccines, and a safer food supply. Others listed specific examples: patents for diagnostics for spider lamb syndrome and for bovine leukemia virus, Bt-corn, and screening procedures for *E. coli* O157 and *Salmonella*. Several respondents were not sure about the answer or did not have enough information to respond. One of the respondents who marked "no" commented that major direct benefits were coming from industry research; another wrote that NRI grants did not directly or immediately result in major benefits.

Nonrecipients:

About a dozen respondents either did not have the information or felt that it was too early to evaluate the NRI's impact on US agriculture. Another group of respondents, similar in size, felt that the NRI had had a great impact by providing the base for the whole biotechnology industry, developing expertise in plant materials, increasing the quality and quantity of fundamental and applied agricultural research, and training young scientists. More specific examples included successfully dealing with major pathogens in cattle, use of bovine somatotropin, and investigative work on the mechanisms by which agricultural pollutants can contaminate the larger environment—which affected congressional action on the types of policies needed to minimize damages, such as the Buffer Strip Initiative (to slow runoff flow and trap sediment from the field by the placement of strips of permanent vegetation).

Land Grant:

Although no specific examples were cited, many comments were made that the NRI contributed to agricultural biotechnology, food-safety management/environmental issues, farming, animal health, and so on. Some respondents felt that there would be no basic research in agriculture without the NRI. Others said that because the program is relatively young, it was too early to see the effects. A comment was also made that not much could be expected from an underfunded program.

Industry:

Most comments were favorable about NRI contributions. General examples included contributions to animal-health monitoring, corn and soybean research, basic R&D, and pork and turkey quality through better understanding of muscle physiology. One person felt that the program was too young to evaluate the results.

8. Are the areas of congressionally mandated funding appropriate? These areas are: plant systems; animal systems; nutrition, food quality and health; natural resources and the environment; processes for adding value and new products; and markets, trade and policy.

	Awardees	Nonrecipients	Land Grant	Industry
Yes	106	38	50	26
No	17	6	7	4

What two areas of research are most important to you, in order of priority (1 = highest)?

	Awardees		Nonrecipients		Land Grant		Industry	
	Rank 1	Rank 2	Rank 1	Rank 2	Rank 1	Rank 2	Rank 1	Rank 2
Plant Systems	28	20	7	3	22	4	7	2
Animal Systems	20	10	10	1	14	5	6	3
Nutrition, Food Quality and Health	14	16	3	11	6	7	5	11
Natural Resources and the Environment	12	17	6	8	11	3	2	1
Processes for Adding Value and New Products	2	4	1	—	2	3	5	5
Markets, Trade and Policy	2	—	—	1	4	3	—	—

Awardees:

A vast majority of respondents agreed that the funding areas were appropriate. Many of comments were made about insufficient funds. Some indicated that the NRI should focus more on long-term basic research. Many would like to see bigger emphasis on their program areas with increased funds and new subareas. There were few comments that the Markets, Trade and Policy, area and the Processes for Adding Value and New Products area were "soft" and should not be funded by the NRI but left to industry or commerce.

Nonrecipients:

A few respondents commented that it was not always clear what the program areas covered. For example, Plant Systems and Natural Resources and Environment could be closely linked, and entomology research could be under both Plant Systems and Animal Systems. Recommendations were made that, for example, Plant Systems and Animal Systems become one program area. Some researchers called for restoring discontinued program areas, such as forest/range/crop/aquatic ecosystems or forestry pest management. Others recommended creating new areas, such as fiber production. One respondent wrote that "detailed program guidelines simply do not allow researchers to span access boundaries in addressing salient questions … current guidelines are too traditional and not receptive to the 'systems' view that producers must employ." Comments were also made that awards are too small.

Land Grant:

Only a few respondents felt that the program areas should not be determined by Congress. Some respondents made specific recommendations, such as increasing funding, moving Value Added/New Products from the NRI to commodity or industry groups, adding Genomics, integrating Plant Systems with Animal Systems, moving Markets, Trade and Policy to the Department of Commerce, and changing the NRI from a discipline to a problem-oriented program.

Industry:

A majority of respondents agreed that the program areas were appropriate but underfunded.

9. If the NRI received a significant increase in appropriations, how should the additional funds be used? Rank the following (1 = highest; 5 = lowest).

Awardees:

	Rank 1	Rank 2	Rank 3	Rank 4	Rank 5	Marked without rank
Expand into new areas	16	14	28	42	16	2
Increase size of awards	60	29	13	9	5	17
Increase duration of awards	25	47	30	11	2	11
Increase number of submission dates per year	21	17	29	30	13	7
Other	5	7	4	2	21	4

An overwhelming number of respondents called for increases in the size and duration of awards, in the number of awards (too many excellent proposals are not awarded), and in the number of submission dates per year (from one to two; especially important when submitting revised proposals).

Nonrecipients:

	Rank 1	Rank 2	Rank 3	Rank 4	Rank 5	Marked without rank
Expand into new areas	11	4	8	6	5	3
Increase size of awards	12	6	7	10	—	4
Increase duration of awards	3	17	9	3	1	4
Increase number of submission dates per year	9	6	5	8	10	8
Other	6	1	1	2	6	10

Major recommendations were to increase the number of awards and to increase funding. Others included increasing funding for increasing funding for specific areas, and awarding riskier research initiatives.

Land Grant:

	Rank 1	Rank 2	Rank 3	Rank 4	Rank 5	Marked without rank
Expand into new areas	7	4	22	8	3	3
Increase size of awards	21	12	6	5	1	4
Increase duration of awards	9	30	9	5	1	2
Increase number of submission dates per year	4	2	9	17	9	3
Other	7	3	1	2	7	1

In addition to increasing the funding and duration of awards, many recommendations were made to increase the number of awards. Other recommendations included diversifying the pool of recipients, expanding into new areas (genomics, cropping systems, and tropical agricultural problems), reducing emphasis on grain and fiber crops and increasing emphasis on plant fruits and vegetables, funding "system-type" projects, changing orientation from areas to problems, and placing emphasis on interdisciplinary research.

Industry:

	Rank 1	Rank 2	Rank 3	Rank 4	Rank 5	Marked without rank
Expand into new areas	3	2	5	9	3	1
Increase size of awards	14	5	2	1	1	1
Increase duration of awards	2	11	6	3	—	1
Increase number of submission dates per year	3	4	8	8	—	—
Other	—	1	3	—	5	1

Recommendations were made to increase the funding, number, and duration of awards. Others included establishing a category designed to encourage industry collaborative efforts (earmarking the percentage of funds for such proposals), expanding into regional research stations, focusing on fewer areas, focusing within the current areas, and focusing on multidisciplinary research.

10. Should the NRI continue? Mark all that apply.

Awardees:

	Very much	Somewhat	Not much
Is the NRI continuance important to you?	127	10	1
Is the NRI continuance important to your institution?	112	19	5
Is the NRI's continuance important for the U.S.?	126	7	—
Is the NRI an important part of the USDA Research portfolio?	120	4	1
Is the overhead rate (now 19%)			
— acceptable? (29 marked without ranking)	40	22	—
— a deterrent? (3 marked without ranking)	5	15	1

The table shows that the NRI program plays a crucial role for individual researchers, their employers, USDA, and US agriculture as a whole. Most of the comments focused on the overhead rate. Although the table show that 19% rate is acceptable to most respondents, many indicated that it was not acceptable to their institutions. As one researcher summarized, "It is good for PI, bad for institution." One complained that some collaborators were unable to participate in their research, because of low overhead. Another indicated that low overhead was "the only thing that makes low awards 'workable'."

Nonrecipients:

	Very much	Somewhat	Not much
Is the NRI continuance important to you	34	8	8
Is the NRI continuance important to your institution?	37	10	3
Is the NRI's continuance important for the U.S.?	37	10	2
Is the NRI an important part of the USDA Research portfolio?	39	8	2
Is the overhead rate (now 19%)			
— acceptable? (13 marked without ranking)	11	9	1
— a deterrent? (1 marked without ranking)	1	2	—

Most of the comments revolved around the overhead rate and were similar to those made by awardees. Some respondents indicated that their institutions encouraged sending applications to NIH or NSF rather than to the NRI. The low overhead rate could be acceptable to institutions if NRI funds were higher.

Land Grant:

	Very much	Somewhat	Not much
Is the NRI continuance important to you?	41	12	4
Is the NRI continuance important to your institution?	43	7	7
Is the NRI's continuance important for the U.S.?	49	6	1
Is the NRI an important part of the USDA Research portfolio?	47	4	2
Is the overhead rate (now 19%)			
— acceptable? (11 marked without ranking)	12	6	1
— a deterrent? (4 marked without ranking)	3	5	—

Comments discussed only the overhead rate. All agreed that 19% was not sufficient and that full rates should be allowed.

Industry:

	Very much	Somewhat	Not much
Is the NRI continuance important to you?	11	8	2
Is the NRI continuance important to your institution?	8	8	5
Is the NRI's continuance important for the U.S.?	14	4	1
Is the NRI an important part of the USDA research portfolio?	15	3	2
Is the overhead rate (now 19%)			
— acceptable? (6 marked without ranking)	2	3	—
— a deterrent?	1	1	—

One respondent wrote that 19% overhead rate was "in line with most institutions", another that it should not increase, and another that the "probably keeps institutions from pushing researchers too hard to apply for NRI grants for unwarranted work."

11. How would you improve the NRI? Mark all that apply.

Change ...	Awardees		Nonrecipients		Land Grant		Industry	
	Yes	No	Yes	No	Yes	No	Yes	No
...program areas	36	89	21	23	15	36	8	7
...Application process	29	94	14	27	11	40	6	8
...Review process	24	96	27	16	19	31	4	12

Awardees:

A majority of respondents voted against changes in program areas, the application process, or the review process. Many respondents emphasized again the importance of increasing the size and duration of awards and the number of awards and submissions per year. Recommended changes in the review process included creating more permanent review panels (for example, 3-year terms with one-third of the panel rotating off each year), more rapid and frequent reviews, banning the use of reviewers competing for the same funds, and ensuring that qualified reviewers are used in the process. Many repeated recommendations to change (extend) their own program areas. Some felt that the program should focus more on basic research, others more on applied.

Nonrecipients:

Many respondents emphasized the need to change the review process. Recommended changes included better selection of panel members and reviewers to avoid "old boys" networks, ensuring that only competent researchers review proposals, ensuring that all types of institutions are represented on panels, limiting the number of USDA researchers in the decision-making process, and creating more permanent panels. Recommendations were also made about the application process. Researchers spend too much time on writing proposals that are not funded. One respondent recommended that ad hoc reviews should be sent to applicants before panels meet and that applicants should be allowed to prepare a response for review by the panel with the reviewers' comments. Others recommended screening preproposals, shortening turnaround time, and increasing the number of submission dates per year. Other recommendations included expanding or adding program areas, developing a better mechanism for handling integrated areas, and awarding more "adventurous" proposals. One respondent suggested creating an exploratory research program for ideas that are not in line with "normal" funding routes. Many recommendations were also made to increase the size, duration, and number of awards.

Land Grant:

Many comments made by these respondents on question were repeated here. In addition to increasing funding duration, and number of awards, recommendations were made regarding the review process, such as diversifying panels by adding constituent groups (users), adding scientists who represent various geographic areas and cover a broader spectrum of science; and adding a relevance review. Some recommended expanding program areas to customer or citizen concerns and priorities, tropical and subtropical plant and animal research, cropping systems, food safety and environment, and epidemiologic issues. Others recommended decreasing the number of program areas and focusing on science. Also, more funding was recommended for land grant universities and small institutions. More innovative projects should be funded, and the funding rate of applied or mission-related projects should be improved.

Industry:

Recommendations were made to increase the size of awards, to implement a preproposal screening process, to specify targeted research areas, to provide a list of experts ("sounding boards") for discussing proposals, to expand into multiple products from single-source feedstocks, to place emphasis on nutrition, to connect programmatic support from the NRI to universities or private laboratories with industrial long-term needs, to screen projects for maintaining US competitiveness, to consider smaller local and regional research organizations for awards, and to separate plant-science programs from commodity lobbies.

INTERVIEWS WITH FORMER CHIEF SCIENTISTS

In October 1998, the committee conducted separate interviews with four former chief scientists: Paul Stumpf, 1989–1991; Arthur Kelman, 1992–1993; James Cook, 1994–1995; and Ronald Phillips, 1996–1997. The committee asked about their roles and experience with the NRI and about recommendations for the future. The chief scientists provided direct insight into the "nuts and bolts" of the NRI and its interactions with USDA and other agencies. Each of them was proud of the NRI program and had a deep personal commitment to its principles and practices. The discussion below summarizes their opinions and advice. There was general agreement on all major issues, although they placed different emphases on some issues.

Major Issue

Each chief scientist strongly believed that the NRI suffers from inadequate funding and that the original target for NRI funding of $500 million per year would provide the nation with valuable research. That amount would allow a doubling of both the number of funded grants and the size of grants. The inability to reach the original level and the plateau of current funding were attributed to the lack of strong USDA support, the absence of a strong commitment to USDA by Congress, and inadequate understanding of the importance of food and fiber research during a period of abundant food production in the United States.

Role of the Chief Scientist

The Chief Scientist has primary responsibility for the scientific oversight of the NRI programs, the staff, and the review panels. As administrator of the NRI, the chief scientist reports to the under secretary of agriculture for research, education, and economics (before 1994, it was the assistant secretary for research and education); to the Administrator of the Cooperative State Research, Education and Extension Service; and to the USDA Board of Directors of the NRI, which includes the administrators of the intramural research agencies. Since 1989, the chief scientist has been a member of the National Academy of Sciences (NAS) and has usually been selected from institutions outside USDA. The former chief scientists expressed the view that fresh leadership is important, as is a need for a balance between experience and new perspectives. The chief scientist must be resistant to political pressures and advocates of special interests. All former NRI chief scientists have been plant biologists, and one was a USDA/Agricultural Research Service (ARS) scientist. The chief scientist usually recommends a successor after consultation with NRI staff and administration. Some chief scientists believed that the stature that comes with membership in NAS facilitated interactions with other agencies and provided greater access to influential people during their tenure.

The chief scientist position is a half-time position. It is a difficult job for a scientist who must maintain an active research program at another institution during his or her tenure. The chief scientists noted that the position is equivalent to a full-time job and requires considerable startup time. Generally, the chief scientist serves for 2 years. It was suggested that the term be lengthened to provide the NRI with the benefit of extended experience. However, it could be difficult to find scientists who can command the confidence of the academic community and who feel able to spend such a long period in Washington as chief scientist for the NRI.

The chief scientist is supported by a permanent staff, including a deputy administrator, who can make decisions when the chief scientist is not present, has good historical insight, and interacts effectively with administrators in other agencies. The chief scientist role often includes development of joint programs with other agencies, such as NSF and DOE. The former chief scientists thought that such cooperation would be increasingly important as the scope of research by other agencies embraces food, fiber, and natural resources.

The interface with special-interest groups is a large part of the chief scientist's activities. The NRI research agenda ranges from biotechnology farming practices, to tropical forestry, family farms, use of energy in agriculture and forestry, food safety, and environmental impact. All those areas have urgent research needs and expectations for support from the NRI. Tensions can arise if research communities believe that they are not adequately represented in decision-making and the sharing of resources.

Program Areas

The chief scientists cited several reasons for the increase in NRI program areas. One is the strong interest of specific commodity or interest groups in specific research programs. It is often easier to obtain money from Congress for a new program than to obtain more money for an established program. In some cases, new programs were created through administrative decisions made by USDA or Congress. New programs reflect flexibility in the research agenda. However, once program areas are established, they are vulnerable to pre-emption by smaller research communities that depend on the NRI as for their primary source of support. The chief scientists suggested that a sunset clause would be useful for new programs. A science advisory council could also review program areas.

Fairness in Evaluation of Submitted Applications

All chief scientists expressed confidence in the peer-review and evaluation process. Although the review process is long and thorough, panel members generally view it as satisfying and fair. The funding rate is low, and funding levels are often minimal. But complaints about unfairness are few. Scientists, panel members, and administrators generally view the process as exceptionally fair. The former chief scientists and the participating scientists also emphasized that service on panels is a great learning experience. Scientists who serve as panel managers gain valuable administrative experience in a high-quality research-evaluation process.

Quality of NRI Research

Chief scientists were convinced that the peer-reviewed competitive process substantially enhances the quality of research. The evaluations are valuable in themselves, but there is increased benefit when scientists prepare for competitive grant review. Until the creation of the competitive grants program, effective competitive peer-reviewed evaluation of grant proposals was not a major part of the nation's funding of agricultural research by USDA. Virtually all food and fiber research was evaluated "in house", typically by CSRS review teams. That procedure did not provide the expert-level, critical review that is needed to assess the quality of specific project proposals. The value of the review process in providing suggestions for improvement in proposals is high.

The NRI has attracted new scientists to food, fiber, and natural-resources research. The program is open to all universities, research institutes, and USDA research laboratories. Thus, it fosters increased efforts to conduct interdisciplinary research. Furthermore, it increases the scientific scope beyond purely agricultural disciplines.

USDA Administration and Management

The former chief scientists believed it essential that the chief scientist report directly to the under secretary of agriculture for research, education, and economics. That was specified in the original job description, but it has not been consistently practiced. The under secretary is rarely available, or the position might be vacant. The administrator of CSREES also has oversight responsibility for the NRI. However, NRI policy decisions often require wide stakeholder input and consultation from USDA management. Such assistance and guidance were expected to come from the Board of Directors, which is composed of administrators of the intramural USDA research agencies (such as ARS, the Economic Research Service, and the Forest Service FS). They are often under pressure, however, to increase funding for their own agencies and might not be well informed about NRI issues.

The chief scientists cited a lack of advocacy for the NRI in USDA. The NRI has had an "orphan" status. It has not been fully accepted by USDA and is often perceived as competing with intramural programs for research funds. Congress has reinforced the perception when NRI funding was taken from other USDA funds and allocated to other uses.

Economic Evaluation of NRI Program Areas

One former chief scientist observed that agricultural economists have not been adequately involved in the process of evaluating agricultural research. The economic aspects of research must be integrated into the NRI evaluation process and more broadly related to both basic and applied research in agriculture.

Relationship with Congress

The chief scientists noted the absence of NRI advocacy in Congress. Political support often is directed to mission-oriented research, despite the important role of basic research in the NRI. The original balance of 80 % of basic research and 20 % of mission oriented has been shifted to 60-40.

The chief scientists believe that it is important to raise the research image of USDA, both outside and within the agency. The broad scope of USDA hampers the perception of it as a science-based research agency. One example is the recently funded NSF program on plant genomics. Congress allocated funds for the plant-genome program to NSF even though USDA has a long and successful history of work in plant genomics. Earlier, USDA joined with NSF to support the *Arabidopsis* genome program. USDA actively promoted work on the plant genome for many years and recently led an interagency group to establish a national initiative on the plant genome. However, Congress awarded the new funding to NSF.

TESTIMONY

The committee devoted a full day to receiving testimony from interested stakeholders. The stakeholder were in the following groups:

- Panel 1—professional societies and nonprofit research institutes: Agricultural Research Institute; Council for Agricultural Science and Technology; Coalition on Funding Agricultural Research Missions; Council on Food, Agricultural and Resource Economics; Federation of American Societies for Experimental Biology; Henry A. Wallace Institute for Alternative Agriculture; and Tri-Societies of America.
- Panel 2—industry: Monsanto; Pioneer Hi-Bred International, Inc; and Gala Design LLC.
- Panel 3—universities and experiment stations: USDA Research, Education, and Economics Mission Area Advisory Board; and Experiment Station Committee on Organization and Policy.
- Panel 4—farm organizations: Animal Agriculture Coalition, American Farm Bureau Federation, American Soybean Association, National Association of Wheat Growers, National Corn Growers Association, and National Cotton Council.
- Panel 5—federal agencies: USDA Research, Education, and Economics Mission Area; Office of Science and Technology Policy; and Office of Management and Budget.

Respondents were asked to comment on the four objectives of the study:

- The quality and value of NRI funded research.
- NRI research priorities within major program areas.
- Complementarity of NRI programs with other USDA research programs, the federal agencies, state and regional programs, and the private sector.
- Future changes in the NRI.

Quality and Value of Research

Testimony from all panels consistently cited the high quality of NRI research and generation of new technologies. A transparent process for evaluating NRI outcomes was lacking. Observers indicated that the NRI was not alone among federal research agencies in this measurement problem.

Panelists considered the NRI a vital, effective component of USDA research. The NRI has filled gaps in food and fiber research and trained future generations of scientists. The peer-review system was considered the best way to assess projects and distribute funds. However, the farm groups cited a lack of NRI identity in the overall research agenda, limited awareness of the research funded by the NRI, and a loss in interactions between the NRI and research users.

Representatives of Pioneer, Monsanto, and Gala Design spoke strongly on behalf of publicly supported research. Each indicated that the products they were selling or anticipated selling were based on important contributions by past publicly funded research.

The USDA representatives suggested that low funding of the NRI, compared with NIH and NSF, placed food and fiber research in a less favored position than biomedical research. They also cited a public perception that the high payoffs from food and fiber research in the past obviated high levels of support in the future. There is a critical need to increase public awareness of the benefits of NRI-supported research. And there is a need for a clearer understanding of the roles of publicly and privately supported research.

Priority-Setting

The panelists consistently cited the need for improvements in priority-setting for NRI research. Although, funding is now "category-driven", most participants supported "issue funding", focusing on solutions to specific problems and issues. Those issues would be translated into specific requests for proposals. Such problem-focused research would be more easily understood by stakeholders and more effectively related to the rest of USDA research and to the research agendas of other federal agencies. Many participants perceived an absence of national goals for food, fiber, and natural-resources research. The Food Animal Integrated Research of 1995 (FAIR '95) and the Coalition for Research on Plant Systems of 1999 (CROPS '99) priority-setting initiatives were useful, but they fell short of achieving national research priorities.

Much discussion also centered on the distribution of funds between basic and applied research. Commodity groups tend to favor applied research; while professional societies emphasize basic research. Striking an effective balance is a persistent challenge.

Industry representatives commented that the NRI priorities are too broad for the available funding. They felt that NRI priorities overemphasized commodity approaches and that commodity groups have enormous influence on funding, as in the case of genomics. The industry panel recommended that areas of interest to food and fiber biotechnology stakeholders funded by the public sector include soil management and carbon sequestration; seed biology (carbon, nitrogen, phosphorus, sulfur, and trace-element mobilization and use); biotic stress tolerance (plant metabolism and heat and cold resistance); pest management and pathogens; analytic tools for studying grains; nutrition (understanding of plant-derived nutrients); molecular breeding tools; and bioinformatics.

At least two panelists noted the relatively low funding of social-science research by the NRI, including how scientific discoveries affect the social and operating structure of the food system. Those interested in social sciences emphasized a systems approach to research in which social-science investigation would play an important part.

The discussion of social sciences also was linked to mission-driven research. Several panelists observed that the NRI lacked a mission orientation despite its statutory mandate. Some pointed out that stronger observance of mission would move the NRI toward applied research and away from basic research. One participant observed that the NRI research niche is not clear either within USDA or with regard to the overall needs of the food and fiber system.

The need for interdisciplinary or integrated research received considerable discussion, especially in light of limitations attributable to the small grant amounts. One participant questioned whether additional funds were needed, because "good researchers today are getting good funding and are working at 150 % of capacity." Others, particularly in the professional societies and universities, differed.

Finally, the concepts of scientifically sound research and "relevance" received long discussion. Participants recommended implementing a two-tier proposal-review process. In tier 1, scientists and industry representatives would jointly review relevance; tier 2 would be a scientific merit review. It became clear during the discussion that relevance could reflect stakeholder understanding and involvement in the NRI process. Everyone wanted scientifically sound research, but some were more interested in short-term applied research to help solve current problems.

Stakeholders

All participants indicated the need for increased transparency in the priority-setting and granting processes of the NRI. The FAIR '95and CROPS '99 initiatives were cited as ways in which all stakeholders can participate in priority-setting. Again, stakeholders closest to the farmer had a more immediate view of their needs, and those closest to the universities and scientific societies had a more basic, long-term approach. To be fair, the commodity groups indicated that they understood the importance of basic research. They want, however, to tell their members how a basic-research concept or idea might ultimately result in the solution of a problem that their members experience. Some commodity groups indicated increasing disenchantment with the NRI because of the lack of transparency and the small amount of available funding.

University panelists also suggested evidence of reduced interest in the NRI because of small grant amounts, short grant duration, low approval rates, and low overhead percentage (14 %). Awardees' institutions must make up the difference between the overhead paid by the sponsor and the standard overhead rate received from a federal source.

All participants agreed that the NRI is not well understood by those who are not principal investigators. Its processes are murky, and its role, especially in research training, is under-appreciated. All participants insisted that increased basic research in food, fiber, and natural resources is absolutely required but, they felt that better priority-setting and the involvement of stakeholders in understanding the direction, purposes, and operations of the NRI were essential to its future success.

Complementarity

All the panelists indicated that the integration of research across disciplines, institutions, and sectors of the economy is greater than it was a decade ago. The need for effective NRI partnerships and the federal research community and other USDA agencies was noted repeatedly.

Internal Processes

Participants responded favorably about the day-to-day operations of the NRI, although feedback to those not receiving NRI grants might be improved.

Should the position of chief scientist be a full-time position or continue as a part-time position? The arguments for full time were based on continuity, priority-setting, and relationships with stakeholders. The arguments against were the "burnout nature" of the job and whether qualified scientists would be willing to serve for more than 2 years.

Committee members often asked whether the NRI would thrive outside the "USDA budget box". The participants believed that it would be politically difficult to achieve and might eliminate NRI funding altogether if it were attempted.

Funding

Almost every participant volunteered that the NRI was dramatically underfunded. Grants were seen as too short, too small, and with too low an overhead allowance. The transaction costs of applying to the NRI were seen as high, given the odds of receiving a grant and the amount of money available. was referred to repeatedly. The original National Research Council study (1989) that recommended $500 million for the NRI. Representatives of federal agencies strongly supported increases in NRI funds. They emphasized, however, that limited funding for the NRI resulted from the limited amount of money available for all agricultural programs (such as research, rural development, soil conservation, and forest service.)

It was not the committee's specific task to investigate the level and disposition of funding for the NRI, but the panelists offered many comments as to why food and fiber research, as an area of national need, has been chronically underfunded during a period of rising resources for both NSF and NIH. Those discussions included lack of money, the inability of users to advocate food and fiber research, internal competition within USDA, lack of support by USDA and university researcher performers, the paucity of "rural" or food system representation in Congress, and competition among various congressional committees to control funds.

Changes for the Future

Participants suggested numerous changes for the NRI. The first was more funding. The second was the development of a consistent and well-understood priority-setting process. The third was coordination of NRI research priorities with those of USDA and other federal agencies. Participants found the idea of setting up six institutes covering the six NRI divisions appealing. The institutes would be problem- and issue-driven rather than category-driven. Another prominent suggestion was to from an external advisory committee to help in setting priorities and relating to stakeholders; this advisory committee could reflect the reorientation of the NRI toward a problem-based priority-setting process.

References to the "Oregon Invests!" accountability system piqued interest (GAO, 1996). The "Oregon Invests!" database has proved to be a reliable source of information about the economic, social, and environmental consequences of agricultural research programs. By providing that accountability in easily and quickly accessible forms, it has helped to stimulate strong legislative support for the research enterprise of the Oregon Agricultural Experiment Station. Participants triggered discussion about how one enumerates and evaluates research outcomes that are essential to continued support.

The testimony helped the committee to understand the quality of NRI-funded research, the importance of measuring that quality, and the need for effective priority-setting the NRI, and the need to address how NRI activities complement USDA and other federal programs. The committee heard thoughtful, feasible suggestions for change, particularly in how the NRI might relate more openly and receptively to current and future stakeholders.

Appendix D

Outline for Interviews (June 16, 1998) with USDA Professional Staff on Internal Workings of the NRI Program

Note 1: Summary material from the first committee meeting was used to craft various questions and statements of issues that should be touched on during the day-long interview discussions with USDA staff. They are offered here as a framework of possible approaches to conversations.

Note 2: A recurrent theme to be probed in the interviews is the difference between expectations raised by the creation of NRI and the perception that it has yet to fulfill them due to congressional limits on budget, less than full participation by segments of the agricultural research community, and the difficulty in changing culture – both within an agency and in the research-performing institutions it supports.

I. Main Categories of Guiding Questions/Issues

A.Funding and Oversight

• Why has this program been so poorly funded (at only 10 percent of original plan)?

• Perceived Congressional "neglect" to NRI: national competitiveness and national security.

• How does the NRI fit in with other grant programs?

B.Internal Processes

• Identify priorities within NRI and explore how they were set.
• Integrity of the peer review process.
• Concern about "good old boy" perception of grant recipients.
• Identify the realities of multidisciplinary research (including social sciences).
• Concern about minimal industry involvement. Question of how to complement peer review process with aspects of industry evaluation.

C.Impacts/Measures of Success

• Is the NRI bringing in new young scientists and fresh research ideas?
• What's the process of evaluating retrospective studies?

II. Issues Arising from USDA Staff Appearance before the Committee

A. What is the role of public input to existing research mechanisms and how do NRI mechanisms help to solve problems effectively?
B. Faced with the current funding limitations, how does NRI need to focus the current program? USDA seeks guidance on overall management (or "portfolio") questions such as number vs. size of awards, program priorities, and program collaboration.
C. What qualifies as performance measures of mission-focused research that is at the same time high-risk, i.e., projects traditionally funded by the NRI?
D. NRI's unique challenge is to advance the basic science that will address issues not even identified yet. How, then, does the program effectively redefine agricultural science research initiatives to be anticipatory?

For the NRI Leadership:

A. **What are the reasons behind NRI's stagnancy?** The traditional agricultural science community was weaned on formula funds to build institutional capacity, not competitive grants to support specific projects. Is the tradition a significant continuing barrier to research constituency participation? To congressional support?

B. **Who is an advocate of the NRI in Congress?** Who is politically a positive force for the NRI –scientific societies, lobbyists, commodity groups (re: the Farm Bill)?

III. Committee Questions To USDA Staff (during April 29 appearance) Embellished for Follow Up in Interviews

For Program Staff:

A. **Are the program and the constituency in conflict? Is the constituency's reluctance due to the size and length of the awards, number of awards, a perceived lack of mission clarity, or other perceptions?** Non-land grant scientists and scientists in nontraditional areas in agriculture are not well represented within the constituency. Public perception of agriculture research and its importance play a big part. How critical is the perception that the NRI is insulated and disconnected from other agricultural science research programs?

B. **Is the quality of the NRI program an issue?** The standard of quality seems to rely on the peer review panels, i.e., success in landing top scientists to serve on the panels. Similarly, research with high consensus peer rankings must be declined each year due to limited funds. Each of 26 research program areas is evaluated by one of about 30 panels (consisting of about 10 scientists). Last year the NRI was able to fund 24% of the proposals received (though another 25% could be funded based on quality).

C. **How can NRI's distinct contribution or value added to the USDA's agricultural research portfolio be determined?** Systematic data on NRI's niche relative to NSF, NIH, and the private sector, are lacking. Service by the same scientists on NRI, NSF, or NIH review panels is one measure. Submission of identical proposals to one or another agency would suggest no perception of uniqueness in program mission. Vital information would be the agency of choice, i.e., the sequence of submission before and after decline.

D. **How much latitude does NRI staff enjoy in focusing program themes and support?** If Congress is micromanaging the program, though it specifies by statute only six broad areas, then should more discretion be delegated to USDA staff, a la NSF and NIH, to refine priorities, set requirements for multi and single disciplinary work, determine the amount of mission linked and fundamental work, as well as a percentage of what types of institutions to fund? Continuity is important for support from the scientific research community, and funding across the six areas has been stable since 1992. In a competitive program, a tension will exist between how directive of and how responsive to the research community the program staff can be.

E. **Should the NRI be more global as opposed to national in its orientation?** Other issues to be addressed are genomics research, food safety issues, and agriculture in the environment. The role of technology in all of this is a key factor, too.

F. Should the distribution of program funds change by type of **performer?** Why or why not? Since 1991 the overall funding profile has been land grants (70%), other colleges and universities (16%), intramural USDA labs (4-5%). NRI is unique in that any federal agency, including USDA, can apply for an award.

IV. Committee Issues to be posed as Questions

Efficiency:

• What evidence can you cite that the NRI process is administratively efficient?

• What is sacrificed in the name of efficiency? (Is this quality overrated?)

A. Responsiveness:

• How do you know that the program is responsive to advances in the agricultural knowledge base?

• What feedback do you get from panel members, proposers, and others to indicate responsiveness (or lack thereof) to community consensus on priorities, funding decisions, etc.?

• Is feedback received largely within a panel context? Give examples.

B. Fairness:

• Fairness is at the core of any peer review process. How do you ensure that proposals are treated even-handedly?

• What do you tell proposers about your process?

• How do panel members help/hinder perceptions of fairness?

• What are the most common complaints about "unfairness"? Do they relate more to process or to outcome?

C. Overhead calculation:

• How much of a deterrent is the 14% overhead cap, i.e., what is lost by its imposition and enforcement?

• What could be done with no cap that is precluded at present?

• Is the cap a lightning rod for other criticisms/complaints about program effectiveness, lack of submission, congressional indifference, etc.?

V. Issues Identified by the Committee as Relevant for USDA Staff Elaboration (esp. Chief Scientists/Past Directors)

A. Administrative

- priority setting (applied vs. basic)
- internal management perceptions
- feedback

B. Political

- political independence
- legislative pressure

C. Niche/Portfolio

- perceptions of connections of program to USDA...etc.
- interagency process

D. Individual Perspective

- leadership influence: what did you initiate, what impact did you have?
- why did you take/leave the job?
- would you do it again?
- perception of length of service (2 years vs. full-time/perm?).

Appendix E

Committee Members' Experience with the National Research Initiative, the National Institutes of Health, and the National Science Foundation

On the basis of information provided by committee members, the committee's experience with the NRI, NIH, and NSF can be summarized as follows:

	No. Committee Members		
Type of Experience/Activity	NRI	NIH	NSF
Panel member[a]	7	4	3
Grant recipient	10	3	6
Application turned down	8	5	6
Application pending	3	0	1
Currently using grant	8	0	4
Never applied	3	8	5

[a] Four committee members indicated participation in panels of other agencies or programs (Department of Energy, National Institute of Standards and Technology, Small Business Innovation Research, and Binational Agricultural Research and Development Foudnation), and four committee members have not served on any panel.

As educators, six committee members supervised about 40 NRI grant applications (while not on NRI grants), of which 20 were successful.

Appendix F

NRI Applications Funded

	New Proposals			Renewals		
		Awarded			Awarded	
Year	No. Submitted	No.	%	No. Submited	No.	%
1995	2,147	422	20	325	150	46
1996	1,969	380	19	345	149	43
1997	1,841	396	22	323	151	47
1998	1,688	401	24	335	157	47
1999	1,932	446	23	263	121	46

Appendix G

NRI Funded Contributions to Major Scientific Advances in Food, Fiber, and Natural Resources

Of the 3,070 proposals submitted to the NRI in 1995–1997, 745 were awarded with a total of $89,637,786 (R. Michael Roberts, NRI, Sept. 1, 1998, personal communication).

This appendix summarizes what NRI staff considered some leading examples of NRI-funded contributions to major scientific advances in food, fiber, and natural resources. The information presented here was supplied to the committee by the NRI staff.

NRI FUNDED RESEARCH THAT HAS LED TO MAJOR SCIENTIFIC ADVANCES

Biologically Safe Protection of Wheat from Take-all Disease by Using Soil Bacteria

Crop rotation is the best-known practice available to farmers for management of the soilborne plant pathogens responsible for potentially devastating root diseases, wilts, stem rots, and blights of crop plants and for such cosmetic diseases as common scab of potato. Growing unrelated crops in a rotation allows time (about 2 years or more) for the soilborne pathogens of a crop to die out or be eliminated by their natural enemies before the same crop is planted again in the same field. However, scientists have known that soil microorganisms can suppress soilborne pathogens after many years of a crop monoculture (that is, without crop rotation). Until recently, US agriculture has not taken advantage of microorganisms to protect crops against harmful pathogens.

David Weller, R. James Cook, Jos Raaijmakers, and Linda S. Thomashow (CRGO/NRI funded scientists) have spent the last 2 decades studying the microbiologic basis of the decline of take-all disease after 12-15 years of wheat monoculture. Take-all is the most serious root disease of wheat worldwide because most of the world's major wheat-growing areas, lacking economically suitable broad-leaf crops that could be grown in alternate years, use little or no crop rotation. Take-all is especially devastating where wheat is planted with reduced tilling to reduce soil erosion, and it has been known to wipe out entire fields.

Soil laden with some strains of root-associated *Pseudomonas*, including *P. fluorescens* and *P. chlororaphis*, appears to suppress the fungal pathogen. These scientists have shown that those bacteria produce natural antibiotics (mainly 2,4-diacetylphloroglucinol and phenazine-1-carboxylate) that stop the growth of the fungus. Their research has provided the first proof that the ability of soil microorganisms to produce antibiotics is critical to their survival and activity in soil and that antibiotic-producing soil microorganisms constitute one of nature's most effective methods for management of plant diseases. In addition to the knowledge of how to manage the beneficial bacteria through the cropping system, these strains of *Pseudomonas* can be grown in industrial quantities in fermentors and applied as a biologically safe seed coating that prevents take-all disease in wheat.

Three patents have been issued, two more are pending, and two license agreements have been issued for use of specific strains in turf, as well as wheat.

In 1997, Weller, Thomashow, Cook, and Raaijmakers received the Ruth Allen Award; this is the highest award for research given by the American Phytopathological Society and recognizes contributions to science that have changed the direction of research.

CRGO and NRI Awards to Cook and colleagues:

- 1978; Biological Stress on Plants; $120,000; 3 years.
- 1981; Biological Stress on Plants; $60,000; 2 years.
- 1986; Biological Stress on Plants; $100,000; 2 years.
- 1989; Plant Pathology/Weed Science; $100,000; 2 years.
- 1991; Plant Pathology/Weed Science; $120,000; 2 years.
- 1991 (1993 renewal); Plant Pathology; $100,000; 2 years.
- 1994; Soils and Soil Biology; $212,000; 3 years.
- 1996; Plant Pathology; $116,041; 2 years.
- 1994 (1997 renewal); Soils and Soil Biology; $255,000; 3 years.

Patents:

• 4,456,684; June 26, 1984; D. Weller and R. J. Cook; Method for screening bacteria and application thereof for field control of diseases caused by *Gaeumannomyces graminis*.

• 5,955,298; September 21, 1999; L. S. Thomashow, M. Bangera, D. Weller, R. J. Cook; Sequences for production of 2,4-diacetylphloroglucinol and methods.

• 5,972,689; January 24, 1997; R. J. Cook, D. Weller, D.-S. Kim, L. S. Thomashow; Methods and compositions for the simultaneous control of the root diseases caused by *Gaeumannomyces graminis*, *Rhizoctonia*, and *Pythium*.

• Filed November 20, 1997; J. Raaijmakers, L. S. Thomashow, D. Weller, R. J. Cook; Biocontrol agents for take-all.

• Filed December 18, 1998; Z. Huang, L. S. Thomashow, D. v. Mavrodi, J. Raaijmakers, D. Weller, R. J. Cook; Transgenic strains for biocontrol of plant root diseases.

Publications:

• Raaijmakers, J. M., D. M. Weller, and L. S. Thomashow. 1997. Frequency of antibiotic producing *Pseudomonas* spp. in natural environments. Applied and Environmental Microbiology 63(3):881–887.

• Bonsall, R. F., D. M Weller, and, L. S. Thomashow. 1997. Quantification of 2,4-dacetylphloroglucinol produced by fluorescent *Pseudomonas* spp. in vitro and in the rhizosphere of wheat. Applied and Environmental Microbiology 63(3):951–955.

• Raaijmakers, J. M., D. M. Weller, R. F. Bonsall, and L. S. Thomashow. 1995. Primers and probes to detect soil pseudomonads that produce 2,4-diacetylphloroglucinol and phenazine-1-carboxylic acid. Phytopathology 85:1191 (abstract).

• Cook, R. J., L. S. Thomashow, D. M. Weller, D. Fuyimoto, M. Mazzola, G. Bangera, and D. Kim. 1995. Molecular mechanisms of defense by rhizobacteria against root disease. Proceedings of the National Academy of Sciences 92:4197–4201.

Decreasing Milk Fever In Dairy Cows: A Major Advance

Milk fever is an important metabolic disorder of dairy cows related to the onset of lactation when cows are unable to maintain normal blood concentrations of calcium. Serum calcium often decreases to a point that does not support normal nerve and muscle function. Cows then suffer severe appetite loss, generalized weakness or collapse, and, if left untreated, death. This disorder affects about 6-8% of all US dairy cows each year, directly costing the dairy industry up to $20 million.

FY 1993, USDA Agricultural Research Service scientists J. P. Goff and R. L. Horst, at the National Animal Disease Center in Ames, Iowa, received an NRI grant to determine whether the potassium, sodium, or calcium concentration in the ration fed to cows just before calving influences their susceptibility to milk fever. At the time of the award, dietary recommendations urged restricting dietary calcium. However, field reports suggested that such a regimen might be unnecessary and possibly even detrimental.

The researchers demonstrated that dietary calcium is not a major risk factor for milk fever. Diets high in potassium or sodium actually induced milk fever by increasing blood alkalinity. Cows on high-potassium diets had lower plasma concentrations of 1,25-dihydroxy vitamin D (important for intestinal calcium absorption) and lower plasma hydroxyproline (an index of the activity of bone calcium resorption). The results suggest that when the blood pH is high, the tissues become resistant to stimulation by parathyroid hormone, a calcium-regulating hormone. The stimulation of bone calcium mobilization and intestinal calcium absorption is then diminished, and normal blood calcium concentration cannot be maintained. Diets high in sodium are not commonly fed to dry cows, but potassium is a cation commonly found in high amounts in the forages included in dry-cow rations.

This work also indicated that the partial success of previously recommended low-calcium diets was due primarily to a reduction in dietary potassium, not calcium.

The expected benefits to the dairy industry are enormous. These findings provide an easily managed feeding approach to the problem of milk fever and already are changing how US cows are fed before calving. The results of this research will decrease the large economic loss due to milk fever. Additional economic benefits will include reductions in the incidences of other important dairy-cattle diseases, such as ketosis and mastitis, for which cows with milk fever are at higher risk.

NRI Award to Goff and Horst:

1993; Milk Fever Risk Factors: Dietary Cation-Anion Differences; $80,000; 3 years.

Publications:

• Goff, J. P., and R. L. Horst. 1997. Effects of the addition of potassium or sodium, but not calcium, to prepartum rations on milk fever in dairy cows. Journal of Dairy Science 80:176–186.

• Popular press publications: Because of the enormous applicability of the findings, the results have been widely disseminated in numerous industry publications, including Hoard's Dairyman; Dairy Today; and Dairy Herd Management.

Storage-Protein Modification in Potato: Tuber Quality and Vaccine Production

With funding initiated in the first year of the USDA Competitive Research Grants Office, William Park (initially at Purdue University, then at Texas A&M University) isolated and characterized the genes for the major potato tuber protein, patatin. Patatin accounts for about 40% of the total soluble protein in potato tubers, but under normal conditions it is not present in large amounts in leaves, stems, or roots. The work led to the development of tuber-specific expression vectors that can be used as a "molecular pickup truck" to direct the expression of any RNA or protein specifically to the economically important part of the plant. It should be noted that the promoters were not patented and thus are freely available in the public domain.

The initial studies set the stage for studies by Dr. Park and others that have advanced understanding of fundamental plant processes, such as the key role of sugars in regulation of plant gene expression and of the mechanism of organ-specific gene evolution. In addition, the resulting potato transformation-propagation system has found practical application in, for example, improving potato cultivars, manipulation of carbohydrate metabolism, and—beyond plant science—production of pharmaceutically important proteins, such as edible vaccines. In the first instance, William R. Belknap and colleagues, at the USDA Agricultural Research Service, Albany, California, are using the patatin-control element as one of several promoters as they seek to develop improved potato cultivars. Preliminary results of plants currently in field trials indicate that the patatin-control element could have use in developing potatoes resistant to bruising damage.

For development of edible vaccines, potatoes expressing a gene encoding the *Escherichia coli* heat-labile enterotoxin B subunit were fed to mice and to humans. In both, ingestion of the transgenic potatoes triggered a mucosal immune response. A second clinical study is now under way to test transgenic potatoes that express another diarrheal antigen, the Norwalk virus capsid protein. The work, by Charles Arntzen and colleagues (Boyce Thompson Institute) with support from the National Institute of Allergy and Infectious Diseases, offers a new strategy in development of safe and inexpensive oral vaccines against human diseases, such as diarrhea and other diseases for which a protective antigen has been defined; tetanus, diphtheria, and hepatitis B are examples. The strategy is being extended to diseases of agriculturally important animals. Avian influenza virus is under investigation by Hugh Mason (also of Boyce Thompson Institute) as a model system for oral vaccines for poultry.

Awards to Park:

- 1981; Regulation of Tuber Protein Synthesis in Potato; $90,000; 2 years; Purdue Research Foundation.
- 1983; Regulation of Tuber Protein Synthesis in Potato; $80,000; 1 year; Texas A& M University.

• 1984; Regulation of Tuber Protein Synthesis in Potato; $73,000; 1 year; Texas A & M University.

• 19856; Regulation of Tuber Protein Synthesis in Potato; $78,000; 1 year; Texas A & M University.

Award to Mason:

1997; Vaccines for Poultry Using Antigens Produced in Transgenic Plants; $100,000; 2 years.

Publications:

• Mignery G. A., C. S. Pikaard, D. J. Hannapel, and W. D. Park. 1984. Isolation and sequence analysis of cDNAs for the major potato tuber protein, patatin. Nucleic Acids Research 12:7987–8000.

• Pikaard C. S., J. S. Brusca, D. J. Hannapal, and W. D. Park. 1987. The two classes of genes for the major potato tuber protein, patatin, are differentially expressed in tubers and roots. Nucleic Acids Research 15:1979–94.

• Bourque J. E., J. C. Miller, and W. D. Park. 1987. Use of an in vitro tuberization system to study tuber protein gene expression. In Vitro Cellular and Developmental Biology 23:381–6.

• Wenzler, H., G. Mignery, L. Fisher, and W. D. Park. 1989. Sucrose-regulated expression of a chimeric potato tuber gene in leaves of transgenic tobacco plants. Plant Molecular Biology 13:347–54.

• Haq, T. A., H. S. Mason, J. D. Clements, and C. J. Arntzen. 1995. Oral immunization with a recombinant antigen produced in transgenic plants. Science 268:714–6.

• Mason, H. S., J. Ball, J. J. Shi, X. Jiang, M. K. Estes, and C. J. Arntzen. 1996. Expression of Norwalk virus capsid protein in transgenic tobacco and potato and its oral immunogenicity in mice. Proceedings of the National Academy of Sciences 93:5335–5340.

• Tacket, C. O., H. S. Mason, G. Losonsky, J. D. Clements, M. M. Levine, and C. J. Arntzen. 1998. Immunogenicity in humans of a recombinant bacterial antigen delivered in a transgenic potato. Nat. Med. 4:607–9.

• Mason, H. S., T. A. Haq, J. D. Clements, and C. J. Arntzen. 1998. Edible vaccine protects mice against Escherichia coli heat-labile enterotoxin (LT): potatoes expressing a synthetic LT-B gene. Vaccine 16:1336–43.

Establishing Nutritional Requirements for Vitamin B6 and Folate in Humans

Vitamin B6, folate, and vitamin B12 function both as coenzymes and as substrates in the one-carbon pathway that is required for the synthesis of nucleic acids and the methylation of DNA. Each can be marginal in the diets of the US population, and direct linkages to public health have been established. Jesse Gregory and colleagues, at the University of Florida, developed technologies to measure bioavailabity, metabolism, and metabolic function of vitamin B6 and folate in rodents and humans. They have provided new information on the bioavailability of dietary folate and daily folate requirements and developed a more quantitative understanding of the metabolic roles of folate and vitamin B6. Major accomplishments have included the following:

- Developed methods based on the use of nonradioactive- (stable-) isotope labeling to assess folate absorption, metabolism, and rate of turnover in humans and animals.
- Developed methods based on stable-isotope labeling to measure the normal rates and effects of folate and vitamin B6 deficiency on key reactions in one-carbon metabolism in humans.
- Demonstrated that the requirement for folate in nonpregnant women is about twice that previously believed. This observation was a major factor in the development of the much higher 1998 recommended dietary allowance for folate.
- Demonstrated that folic acid added to cereal-grain foods (breads, pasta, and rice) is effectively absorbed. The data proved that the recently adopted fortification program is effective in delivering available folate to the US population.

CRGO and NRI Awards to Gregory:

- 1979; Food Composition and Processing Effects on Vitamin B6 Bioavailability; $100,000.
- 1981; The Bioavailability of Folacin in Foods; $100,000.
- 1983; Determination of Vitamin B6 and Folacin Bioavailability Using Isotopic Enrichment Methods; $135,000.
- 1986; Stable-Isotopic and Enzymatic Investigation of Folacin Bioavailability; $120,000.
- 1987; Stable-Isotopic and Radioisotopic Investigation of Folate Bioavailability; $175,988.
- 1991; Stable-Isotope Investigation of Folate Bioavailability and Nutritional Status; $166,842.
- 1992; Folate Nutritional Status and In Vivo Kinetics; $222,521.
- 1994; The Bioavailability of Folate in Foods; $227,618.
- 1996; Folate and Vitamin B6 Dependence of One-Carbon Metabolism; $198,586.

Production of Cattle and Swine Embryos in Vitro, a Prelude to Cloning

Research in the 1960s and 70s developed procedures for test-tube (in vitro) fertilization (IVF) of eggs from some mammals, including the mouse and human. The achievement was illustrated most dramatically by the birth of the first IVF baby, Louise Brown, in 1977. Neal First's laboratory at the University of Wisconsin pursued development of the procedures in livestock to allow immature eggs, obtained from ovaries collected from the slaughterhouse, to provide a vast source of embryos. The general approach was and continues to be collection of ovaries from the slaughterhouse, transportation of ovaries to the laboratory, aspiration of large numbers of oocytes (eggs) from many immature ovarian follicles, and finally in vitro maturation of the oocytes in special culture medium that mimics the conditions involved in egg maturation in the intact animal. There were other technical difficulties to overcome. Tissue-culture conditions had to be developed that allowed sperm from frozen semen to fertilize the matured oocytes and that permitted a fertilized egg to develop into a multicelled embryo. First and colleagues developed the basic methods to produce hundreds of embryos from slaughterhouse-derived material. The spinoff from this work has been enormous, including the following:

- It permitted the cloning of cattle by nuclear transfer from early-cleavage-stage embryos and was a prelude to the eventual demonstration of cloning by somatic cell nuclear transfer. The First laboratory holds the first patent in this field.
- It led to parallel developments in embryo manipulation and cloning in swine.
- It led to the improvement of transgenic techniques for cattle.
- The availability of many embryos has allowed gene expression to be studied during early development of livestock.

For his achievements, First was awarded the von Humboldt Award for Agriculture in 1987 and the Wolf Prize (recognized as the equivalent of a Nobel Prize in agriculture) in 1997.

CRGO and NRI Awards to First:

- 1985; Ontogeny and Control of Development of Bovine Preimplantation Embryos; $161,500.
- 1987; Ontogeny and Control of Development of Bovine Preimplantation Embryos; $142,500.
- 1988; Cellular Regulation of Meiotic State in Bovine Oocytes; $144,000.
- 1988; Development of Porcine Embryos After Multiplication by Nuclear Transplantation; $200,000.
- 1990; Cellular Regulation of Meiotic State in Bovine Oocytes; $70,000.

• 1992; Ontogeny and Control of Development of Bovine Preimplantation Embryos; $213,753.

• 1995; Effect of Bovine Embryonic Stem Cell Origin and Culture on Pluripotency and Totipotency; $120,000.

Patent:

4,994,384; February 19, 1991; R. S. Prather, F. Barnes, J. Robl, N. L. First and V. F. Simmon; Multiplying bovine embryos.

Corn from Cells, Not from Seeds

When the Competitive Research Grants Office was formed in 1978, a grant was awarded to Ronald L. Phillips, of the first Genetic Mechanisms Program Panel, for tissue culture of corn. Phillips had already demonstrated that whole corn plants could be regenerated from cells in tissue culture, and this technology had been quickly adapted to other cereals. Most important, Phillips's CRGO-supported research eventually allowed new genes to be introduced into corn. Before this breakthrough was achieved, however, a major problem had to be solved. Phillips recognized that the plants derived from tissue culture were not exact copies of each other (clones), but that variation was being induced by the tissue-culture procedure. On the one hand, this provided a way of producing new corn varieties; on the other hand, it was a drawback to the production of transgenic plants because the gene could not be introduced into a stable genetic background. The variability was due in part to the activation of mobile genetic elements during tissue culture and to alterations in the methylation of DNA. To circumvent the problem, Phillips used short culture times and then crossed the resulting plants to elite lines.

Genetically engineered corn, particularly varieties that are resistant to insect predation, accounted for more than one-third of the crop in 1998, and its use is increasing. The most common genotype used by industry for producing transgenic crops traces back to Phillips's research supported by CRGO. Tissue-culture regeneration technology of cereal crops has been a major tool in developing genetically engineered plants.

CRGO and NRI grant awards to Phillips:

• 1978; Tissue Culture Genetic Systems in Corn; $110,000; 3 years.

• 1988; Tissue Culture Genetic Systems.

• 1992; Mapping and Isolation of Genomic Regions: Controlling Maturity in Maize; 2 years.

Cloning of Chitinase and Its Practical Application as a Biopesticide

Karl J. Kramer, biochemist with the USDA Agriculture Research Service Grain Marketing and Production Research Center, Manhattan, Kansas, isolated the gene that codes for the chitinase enzyme from the tobacco hornworm *Manduca sexta*. Kramer and collaborators at Kansas State University—S. Muthukrishnan, Lowell Johnson, and Frank White—have cloned and incorporated the gene into tobacco and rice plants. They have demonstrated that coupling the chitinase gene with the bacterium *Bacillus thuringiensis* in genetically engineered plants can be a safe and effective biopesticide that substantially reduces insect growth and curtails foliar damage in plants. Chitinase works by breaking down chitin, a key component in membranes of insect skin and gut tissue. Without this chitin membrane, insects rapidly succumb to microbial infections. The chitinase must be ingested, so it poses a minimal threat to nontarget species because these organisms would have to chew on engineered plants. Because these species are not leaf feeders, they should not even contact the enzyme. Also, chitin is peculiar to insects and other invertebrates and not found in higher animals, so it is considered safe for humans and other mammals. Kramer and collaborators have received continuous USDA competitive grant funding since 1988 to support this research and have published at least 40 papers in scientific journals since the inception of the work. Most recently, they have patented the only known insect chitinase gene used in transgenic plants.

Currently, NRI is supporting Kramer's efforts to transform other plants, such as corn, wheat, and sorghum. Biotechnology companies—including Pioneer, Dekalb, and Prodigene—are working with these scientists to develop practical applications.

CRGO and NRI Awards to Kramer:

- 1996; Development of Insect Chitinolytic Enzymes as Biopesticides; 2 years.
- 1998; Improvement of Insect Chitinase as a Biopesticide in Transgenic Plants; 3 years.

Patent:

5,866,788; February 2, 1999; K. J. Kramer, S. Muthukrishnan, H. K. Choi, L. Corpuz, and B. Gopalakrishnan; Recombinant chitinase and use thereof as a biocide.

Publications:

• Kramer, K. J., A. M. Christensen, T. D. Morgan, J. Schaefer, T. H. Czapla, and T. L. Hopkins. 1991. Analysis of cockroach oothecae and exuviae by solid-state 13C-NMR spectroscopy. Insect Biochemistry 21:149–56.

• Corpuz, L., H. Choi, S. Muthukrishnan, and K. J. Kramer. 1991. Sequence of the cDNAs and expression of the genes encoding methionine-rich storage proteins of *Manduca sexta*. Insect Biochemistry 21:265–276.

• Christenson, A. M., J. Schaefer, K. J. Kramer, T. D. Morgan, and T. L. Hopkins. 1991. Detection of cross-links in insect cuticle by REFOR NMR Journal of the American Chemical Society 113:6799–6802.

• Hopkins, T. L., and K. J. Kramer. 1992. Insect cuticle sclerotization. Annual Review of Entomology 37:273–302.

• Kramer, K. J., L. Corpuz, H. Choi, and S. Muthukrishnan. 1993. Sequence of a cDNA and expression of the genes encoding epidermal and gut chitinases of Manduca sexta. Insect Biochemistry and Molecular Biology 23:691–701.

• Gopalakrishnan, B., S. Muthukrishnan, and K. J. Kramer. 1995. Baculovirus mediated expression of a *Manduca sexta* chitinase gene: Properties of the recombinant protein. Insect Biochemistry and Molecular Biology 25:255–265.

• Zen, K. C., H. K. Choi, K. Nandegama, S. Muthukrishnan, and K. J. Kramer. 1996. Cloning, expression and hormonal regulation of an insect N-acetylglucosaminidase gene. Insect Biochemistry and Molecular Biology 26:435–444.

• Wang, X., X. Ding, B. Gopalakrishnan, T. D. Morgan, L. Johnson, F. White, S. Muthukrishnan, and K. J. Kramer. 1996. Characterization of a 46-kDa insect chitinase from transgenic tobacco. Insect Biochemistry and Molecular Biology 26:1055–1064.

• Choi, H. K., K. Choi, K. J. Kramer, and S. Muthukrishnan. 1997. Isolation and characterization of a genomic clone for the gene of an insect molting enzyme, chitinase. Insect Biochemistry and Molecular Biology 27:37–47.

Additional Information:

A virtual article in Discovery Channel Online is available at http://www.discovery.com/news/archive/news990226/brief5.html?ct=36dafe7f.

Dogs and Infectious Abortion in Cows: A Mystery Solved

Neospora caninum is a protozoan parasite that is a major cause of infectious abortion in dairy cattle. The parasite is found throughout the United States and around the world. In California alone, *Neospora caninum* is diagnosed in 40% of aborted fetuses and costs dairy producers at least $35 million a year. This parasite was first described about 10 years ago, but until last year veterinarians, dairy farmers, and researchers were frustrated by the disease because it was not known how it was transmitted.

A major advance recently occurred in the battle to protect the dairy industry from this parasite. A team led by NRI-supported researcher Milton McAllister, at the University of Wyoming, in collaboration with researchers at Virginia Polytechnic Institute and State University and the Agricultural Research Service, demonstrated that the dog is a definitive host of *Neospora*. This discovery has paved the way for development of effective methods of disease prevention.

Unraveling the mystery of *Neospora* transmission began with a 2-year seed grant in which McAllister and colleagues developed a protocol that would reliably induce formation of large numbers of encysted bradyzoites of *Neospora* in the brains of mice, gerbils, and sheep. The best results were obtained with immunosuppressed mice from which bradyzoites could be purified.

Neospora has structural and genetic similarities to other protozoan species that are transmitted in the feces of carnivorous animals, and results from the 1994 NRI seed grant allowed the research team to obtain a small grant from the American Veterinary Medical Foundation to determine whether cats are definitive hosts; the results were negative. However, as a result of a later award from the NRI's Animal Health and Well-Being Program, dogs were induced to pass the oocysts after being fed the tissues of mice infected with *Neospora* from both canine and bovine isolates.

Scientific understanding of this parasite's life cycle now points to practical ways to decrease its impact. Dairy cows are usually fed rations consisting of feed mixed in large batches. Feed is often piled on the ground in open areas during this process. If a dog shedding *Neospora* contaminates ration ingredients, an entire herd can be exposed to the parasite. The risk of *Neospora* abortion could be greatly reduced by simply maintaining animal feed in a fenced area or closed container. Fencing could also be used to prevent pregnant cows from eating feces-contaminated pasture. Thus, relatively simple methods can be used to solve a multimillion-dollar problem.

The research team successfully completed its stated goal of discovering a definitive host for *Neospora caninum*. The information is being widely disseminated in the popular press (Multiple press releases, Colorado Dairy News, Hoard's Dairyman, Parasitology Today, and so on) . A valuable, new bovine isolate of *Neospora* that was also identified is now available to the research community for further investigations to diminish the impact of this infection. Further studies are needed to determine whether animals related to dogs, such as foxes and coyotes, also transmit the organism. Vaccine development would aid the dairy industry, particularly in management situations where eliminating working-dog and family-pet access to pasture and feed areas might not be possible.

This research success highlights the importance of the NRI's Strengthening Awards Program. Competition for these awards is open to faculty of small and medium academic institutions and at institutions in the USDA Experimental Program for Stimulating Competitive Research.

NRI Awards to McAllister:

- 1994; Induction of infectious cysts of *Neospora caninum* in rodents and fetal lambs; $45,000; 2 year Seed Grant.
- 1996; Determination of the Definitive Host of Neospora caninum; $111,727; 1 year Standard Research Grant.

Publications:

- McAllister, M. M., S. F. Parmley, L. M. Weiss, V. J. Welch, and A. M. McGuire. 1996. An immunohistochemical method for detecting bradyzoite antigen (BAG5) in Toxoplasma gondii-infected tissues cross-reacts with a *Neospora caninum* bradyzoite antigen. Journal of Parasitology 82:354–355.
- McAllister, M. M., A. M. McGuire, W. R. Jolley, D. S. Lindsay, A. J. Trees, and R. H. Stobart. 1996. Experimental neosporosis in pregnant ewes and their offspring. Veterinary Pathology 33:647–655.
- McAllister, M. M., E. M. Huffman, S. K. Hietala, P. A. Conrad, M. L. Anderson, and M. Salman. 1996. Evidence suggesting a point source exposure in an outbreak of bovine abortion due to neosporosis. Journal of Veterinary Diagnostic Investigation 8:355–357.
- McGuire, A. M., M. M. McAllister, W. R. Jolley, and R. C. Anderson-Sprecher. 1997. A protocol for the production of *Neospora caninum* tissue cysts in mice. Journal of Parasitology 83:647–651.
- McGuire, A. M., M. M. McAllister, and W. R. Jolley. 1997. Separation and cryopreservation of *Neospora caninum* tissue cysts from murine brain. Journal of Parasitology 83:319–321.
- Dubey, J. P., M. C. Jenkins, D. S. Adams, M. M. McAllister, R. Anderson-Sprecher, T. V. Baszler, O. C. H. Kwok, N. C. Lally, C. Bjorkman, and A. Uggla. 1997. Antibody responses of cows during an outbreak of neosporosis evaluated by indirect fluorescent antibody test and different enzyme linked immunosorbent assays. Journal of Parasitology 83:1063–1069.
- McAllister, M. M., J. P. Dubey, D. S. Lindsay, W. R. Jolley, R. A. Wills, and A. M. McGuire. 1998. Dogs are definitive hosts of *Neospora caninum*. International Journal for Parasitology 28:1473–1478.
- McAllister, M. M., W. R. Jolley, R. A. Wills, D. S. Lindsay, A. M. McGuire, and J. D. Tranas. 1998. Oral inoculation of cats with tissue cysts of *Neospora caninum*. American Journal of Veterinary Research 59:441–444.

NRI FUNDED RESEARCH WITH HIGH POTENTIAL FOR FUTURE SCIENTIFIC ADVANCES

A Step Closer to Bioengineering Cold-Tolerant Plants

Michael Thomashow and colleagues at Michigan State University used NRI funding from the Plant Responses to the Environment Program to create a cold-tolerant strain of *Arabidopsis thaliana* (a model plant organism). Researchers have known about cold-tolerant genes for close to 30 years, but they had not been able to bioengineer a cold-tolerant plant, because of the number of genes involved. At least 25 genes are associated with cold tolerance, and stable transfer of many genes is not yet possible. However, Thomashow was part of the research teams that discovered that the expression of these genes is regulated by proteins known as transcription factors. In theory, transcription factors can be used to enhance cold tolerance by turning on all the genes involved at one time. Thomashow tried to insert such a transcription factor into *Arabidopsis* and found that, indeed, cold tolerance was enhanced dramatically. Although it is true that the same cold-tolerant genes and a single transcription factor might not work in plants other than *Arabidopsis*, this is a first step in understanding how to control cold tolerance in plants.

NRI Awards (directly related to this research):

- 1988; Plant Responses to the Environment Program; $140,000; 2 years.
- 1990; Plant Responses to the Environment Program; $110,000; 2 years.
- 1992; Plant Responses to the Environment Program; $160,000; 3 years.
- 1996; Plant Responses to the Environment Program; $243,393; 3 years.

Publications (directly related to this research):

• Jaglo-Ottosen, K. R., S. J. Gilmour, D. G. Zarka, O. Schabenberger, and M. F. Thomashow. 1998. Arabidopsis CBF1 overexpression induces COR genes and enhances freezing tolerance. Science 280:104–106.

• Gilmour, S. J., D. G. Zarka, E. J. Stockinger, M. P. Salazar, J. M. Houghton, and M. F. Thomashow. 1998. Low temperature regulation of the Arabidopsis CBF family of AP2 transcriptional activators as an early step in cold-induced COR gene expression. Plant Journal 16:433–442.

Reducing Processing Waste and Negative Environmental Impacts by Wood Modification

Lignin, the complex polymer that cements cellulose fibers together and provides strength and protection for growing trees, is removed when wood is processed for commercial purposes. The removal process requires chemical treatments, and the chemicals and unseparated fibers become a waste product. Ronald Sederoff and other NRI-funded scientists at North Carolina State University have been working on modifying lignin composition in loblolly pine (a commercially important timber source) so that its mechanical strength is reduced and processing efficiency is increased. These researchers discovered that a mutant that produces a modified lignin exists naturally in loblolly pine populations. The mutant appears to change the composition and structure of lignin by blocking the enzyme, cinnamyl alcohol dehydrogenase. The enzyme converts cinnamyl aldehydes to cinnamyl alcohols, which are the precursors of lignin formation. These researchers also showed that laccase was another important enzyme involved in lignin formation, providing yet another enzyme that could be used to manipulate lignin formation genetically.

NRI Grants (directly related to this research):

- 1988; Genetic Mechanisms/Plant Science Bio-Technology (joint); $66,000; 2 years.
- 1991; Plant Growth and Development Program; $110,000; 2 years.
- 1997; Wood Utilization Program; $111,000; 2 years.

Patent:

5,824,842; October 20, 1998. J. MacKay, D. O'Malley, R. Whetten, R. Sederoff. Methods of providing and breeding trees having more easily extractable lignin due to the presence of a cinnamyl alcohol dehydrogenase (CAD) null gene.

Publications (directly related to this research):

- O'Malley, D. M., S. Porter, and R. R. Sederoff. 1992. Purification and characterization of Cinnamyl Alcohol Dehydrogenase in Loblolly Pine. Plant Physiology 98:1364–1371.
- Bao, W., D. M. O'Malley, R. Whetten, and R. R. Sederoff. 1993. A laccase associated with lignification. Science 260:672–674.
- O'Malley, D., R. Whetten, R. Bao, C. L., Chen, and R. R. Sederoff. 1993. The role of laccase in lignification. The Plant Journal 4:751–757.
- Whetten, R. and R. Sederoff. 1995. Lignin biosynthesis. Plant Cell 7:1001–1013.

Bioengineered Safe Resistance to Glyphosate Herbicide

Weed control often requires the use of herbicides that can harm crop plants. Glyphosate (Roundup) is a common herbicide that kills grasses, sedges, and broad-leaf plants by blocking the biochemical pathway that produces essential amino acids (phenylalanine, tyrosine, and tryptophan). Only plants, fungi, and bacteria can make these essential amino acids. Animals (including humans) are insensitive to glyphosate, so its use is relatively safe. Plant resistance to glyphosate has already been genetically engineered with genetic material in the cell nucleus. However, there is now concern over the use of such plants because the resistant genes could be spread with the release of pollen. The release could lead to a decrease in the overall effectiveness of the herbicide against weeds and create "superweeds". Henry Daniell and colleagues, at Auburn University, with the support of NRI funding have found a solution to the problem by using genetic material in the chloroplast to genetically engineer glyphosate-resistant tobacco. Chloroplast genetic material is maternally inherited and cannot be spread by pollen in most crops (with rare exceptions, such as pines). Chloroplast-derived resistance is also more resistant to glyphosate than nucleus-derived resistance. Application of glyphosate after crop emergence is now possible without the fear of uncontrolled spread of the resistance gene or herbicide damage to the crop.

NRI Awards (directly related to this research):

- 1993; Plants Division; $10,648; 2 year Seed Grant.
- 1995; Value Added Products; $120,000; 2 years.
- 1997; Non-Food Characterization/Process/Product Program; $219,438; 3 years.
- 1998; Plant Genome; $160,000; 2 years.

Publication (directly related to this research):

Daniell, H., R. Datta, S. Varma, S. Gray, and S. B. Lee. 1998. Containment of herbicide resistance through genetic engineering of the chloroplast genome. Nature Biotechnology 16:345–348.

Establishment of the Role of Steroid Hormones in Plant Growth and Development

Steroid hormones are crucial for embryonic development and adult homeostasis in animals. Similarly, the insect steroid hormone ecdysone controls many developmental processes in insects. In plants, many steroids have been identified, but only one class of steroids, collectively called brassinosteroids

(BRs), has wide distribution throughout the plant kingdom and unique growth-promoting properties when applied exogenously. Useful agricultural applications of these compounds have been found, such as increasing yield and improving stress resistance of several major crop plants. Despite extensive research, mostly in the 1970s and 1980s, definitive proof that BRs are essential for normal plant growth had been lacking until recently.

It is surprising that a clear role of BRs in plant growth and development came from genetic studies on photomorphogenesis in the model plant *Arabidopsis*, led by Joanne Chory, of the Salk Institute. A mutation in the *DET2* gene resulted in plants that do not respond properly to light. The gene was cloned and shown to encode an enzyme (steroid 5α-reductase) involved in the biosynthesis of BRs. The mutant plant can be rescued by application of BRs. The identification of additional BR-deficient dwarf mutants in *Arabidopsis* and other plant species confirms the importance of these steroids in plant development.

This work is important because it confirms the role of BRs as a major class of plant-growth regulators. Previously, only auxins, gibberellins, abscisic acid, ethylene, and cytokinins were considered "real" hormones, and BRs received little or no attention in botany textbooks or in general reviews of plant physiology and development. Now, because of the work of Chory and others, BRs are receiving a great deal of international attention. It is likely that greater understanding of the molecular mechanisms of BR action could have practical effects on the generation of transgenic crop plants of many species with altered growth properties (such as stature and yield).

NRI Awards:

- 1993; Molecular and genetic analysis of arabidopsis Det 2 gene; 3 years.
- 1996; Molecular and genetic analysis of arabidopsis Det 2 gene; 3 years.

Patents:

- Filed April 1996; Novel plant steroid 5α-reductase, DET2.
- Filed June 1997; Receptor kinase, Bin 1.

Publications:

• Li, J., P. Nagapal, V. Vitart, T. McMorris, and J. Chory. 1996. A role for brassinosteroids in light-dependent development of Arabidopsis. Science 272:398–401.

• Li, J., M. Biswas, A. Chao, D. Russell, and J. Chory. 1997. Conservation of function between mammalian and plant steroid 5α-reductases. PNAS 94:3554–3559.

• Li, J. and J. Chory. 1997. A putative leucine-rich repeat receptor kinase involved in brassinosteroid signal transduction. Cell 90:929–938.

• Fujioka, S., et al. 1997. Arabidopsis DET 2 mutant is blocked early in brassinosteroid biosynthesis. Plant Cell 9:1951–1962.

Additional Information:

Chory has been continuously funded by the NRI since 1991. She has also received funding from the National Institutes of Health, the National Science Foundation, the Department of Energy, and the Howard Hughes Medical Institute for studies on light-regulated gene expression and light signal transduction.

Signal-Transduction Pathway of the Plant Hormone Ethylene

The gaseous plant hormone ethylene has profound effects on plant growth and development. There are numerous responses to ethylene throughout the life cycle of the plant, including induction of ripening in climacteric fruits, promotion of seed germination, promotion or inhibition of flowering, abscission of various organs, and senescence. While she was a postdoctoral researcher in Eliot Meyerowitz's laboratory at the California Institute of Technology, Caren Chang cloned the ETR1 gene from *Arabidopsis*, which was shown to be an ethylene receptor sharing the same "two-component" feature (a sensor and an associated response regulator) of bacterial regulators. It was the first plant hormone receptor to be cloned. Since moving to the University of Maryland in 1994, Chang has continued to work on the molecular and genetic mechanisms of ethylene signal transduction. Important discoveries include the previously undescribed association of a two-component receptor (ETR1) and the MAPK signaling cascade and the repression of responses by the receptors in the absence of ethylene (contrary to the typical signaling paradigms established in animals).

This work is important not only because of the cloning of the first plant hormone receptor, but because it has made the ethylene response pathway one of the best understood signaling pathways in plants. It serves as a model for the study of other plant hormones. It also shows that plants do not necessarily transduce signals in the same way as other eukaryotes. Because numerous physiologic processes are mediated by ethylene, the manipulation of ethylene response is vital to the storage, transport, disease protection, appearance, and flavor of numerous plant products. The isolation of new ethylene signaling components will lead to new strategies for manipulating a variety of ethylene responses and thus help us to refine our ability to control plant growth processes.

NRI Awards:

- 1995; Genetic Dissection of Ethylene Signal Transduction Based on the Arabidopsis ETR 1 Gene; 3 years.
- 1998; Arabidopsis Ethylene Signal Transduction: New Interacting Components; 3 years.

Patents:

5,689,055; November, 18. 1997. E. M. Meyerwitz, C. Chang, A B. Bleecker. Plants having modified responses to ethylene.

Publications:

- Chang, C. and R. C. Stewart. 1998. The two-component system: regulation of diverse signaling pathway in prokaryotes and eukaryotes. Plant Physiology 117:723–731.
- Clark, K. L., P. B. Larsen, X. Wang, and C. Chang. 1998. Association of the Arabidopsis CTR1 Raf-like kinase with the ETR1 and ERS ethylene receptors. Proceedings of the National Academy of Sciences 95:5401–5406.
- Chang, C. 1996. The ethylene signal transduction pathway in Arabidopsis: an emerging paradigm? Trends in Biochemical Science 21:129–133.

Additional Information:

Chang has been supported by the NRI since 1995. The NRI has been her sole source of federal support beyond her postdoctoral fellowship.

Improvements in Plant Biotechnology by Manipulation of Gene Silencing

The ability to use plants as "factories" to produce large quantities of valuable proteins is one of the most exciting and potentially useful developments to come out of modern biotechnology. Transgenic plants might be used to produce medically or industrially valuable proteins in a way that is less expensive, more environmentally friendly, and less reliant on the use of animals. In addition, growing such transgenic plants might provide improved economic opportunities for farmers.

Until recently, the ability to use plants to produce large quantities of proteins has been hampered by the existence of a process in plants called gene silencing, which normally protects the plants from viral pathogens. As the name implies, gene silencing stops the production of proteins that would otherwise be produced at very high levels, such as viral proteins or proteins produced by "transgenes" in plants grown as protein factories.

In 1995, the NRI Plant Pathology Program awarded $149,000 to Vicki Vance, at the University of South Carolina (award 95-37303-1815), for her work aimed at understanding the serious diseases caused by the interaction of two viruses in the same host plant (viral synergism). The award was renewed for $150,000 in 1997. In 1995, the NRI's Plant Pathology Program also made an award to James Carrington, at Texas A & M, for $219,000 for his work on the intercellular movement of potyviruses (award 95-37303-1867). Using materials generated by Carrington's group, Vance's group showed that a protein encoded by one virus permitted a wide variety of unrelated viruses to accumulate to high levels and cause serious disease. They reasoned that the viral protein exerts this effect by paralyzing the host plant's gene-silencing process (Pruss et al. 1997). Papers published independently by Vance's group, and Carrington's group, and a group in England (Anandalakshmi et al.1998; Kasschau and Carrington 1998; Brigneti et al. 1998) provided experimental data to show that this was indeed the case.

Because of the practical implications of this work, Vance and her colleague at the University of South Carolina, Gail Pruss, made a successful US patent application through the University of South Carolina (application 08/827,575, "Method for enhancing expression of a foreign or endogenous gene product in plants"). The coinventors listed on the patent application were Laszlo Marton, Carrington, and William Dawson. Numerous companies have expressed interest in licensing the patent.

Work aimed at understanding how the viral protein interferes with gene silencing continues to be supported through the NRI (award 98-35301-6078 made to Vance by the NRI's Plant Genetic Mechanisms Program and award 98-35303-6485 made to Carrington by the NRI's Plant Pathology Program).

NRI Awards to Vicki Vance, University of South Carolina:

- 1995; Mechanisms of Plant Viral Synergism; 4 years.
- 1998; A Viral Suppression of Gene Silencing in Plants; 2 years.

NRI Awards to James Carrington, Texas A&M University

- 1995; Intercellular Movement of Potyviruses; 2 years.
- 1998; Host Responses to Potyviruses; 3 years.

Patent:

5,939,541; August 17, 1999; V. B. Vance, G. J. Pruss, W. O. Dawson, J. Carrington, M. Laszlo; Method for enhancing expression of a foreign or endogenous gene product in plants.

Publications:

• Anandalakshmi, R., G. P. Pruss, X. Ge,., R. Marathe, A. Mallory, T. H. Smith, and, V. B. Vance. 1998. A viral suppressor of gene silencing in plants. Proceedings of the National Academy of Sciences 95:13079–13084.

• Brigneti, G., O. Voinnet, W. X. Li, L. H. Ji, S. W. Ding, and D. C. Baulcombe. 1998. Viral pathogenicity determinants are suppressor of transgene silencing in *Nicotiana benthamiana*. The European Molecular Biology Organization Journal 17:6739–6746.

• Kasschau, K. D., and J. C Carrington. 1998. A counterdefensive strategy of plant viruses: suppression of posttranscriptional gene silencing. Cell 95:461–470.

• Pruss, G., X. Ge, X. M. Shi, J. C. Carrington, and V. B. Vance. 1997. Plant viral synergism: the polyviral genome encodes a broad-range pathogenicity enhancer that transactivates replication of heterologous viruses. Plant Cell 9:859–868.

A New BAC Library Vector for Transferring Large DNA Inserts to Plants

Numerous potential improvements in the characteristics of agriculturally valuable plants are possible through identification of a desired trait in another organism and its expression in a crop plant. That is accomplished by locating the genes responsible for the trait and inserting them into the crop plant's chromosomes. Some desirable genes, such as those for disease resistance in plants, can occur in clusters, creating the need for a reliable system for transforming large segments (>100 kb) of DNA into plants. The relatively large genomes of plants and the abundance of repetitive DNA sequences in them add to the need for an improved transformation system.

To provide a solution to those long-standing problems in basic plant biology, Carol Hamilton, while at Cornell University, developed a binary-bacterial artificial chromosome technology, called BIBAC, to facilitate the development of new elite varieties of agronomic crops. BIBAC technology not only accelerates the identification of agriculturally important genes, but also makes it possible to introduce valuable traits of interest into plants without dragging along deleterious traits—a common problem for classical plant breeders. The success of the BIBAC system would not have been possible without enhancing the ability of *Agrobacterium tumefaciens* to effect DNA transfer to the plant chromosomes. That is, this work included basic research that has affected the plant-transformation community. Since the development of BIBAC technology, several groups have requested the virulence helper plasmids that made technology a success, not because they needed BIBAC technology itself, but because they were interested in improving the transformation efficiency for their plant system of interest. In general, this is a more common problem for agronomic crops than for model plants used for basic research. The critical elements of the new technology, the bacterial strains and plasmids, have been requested by hundreds of academic and industrial laboratories around the world. In addition, various parties have expressed interest in licensing BIBAC technology; these queries are directed to the Cornell Research Foundation. A US patent has been issued for the BIBAC vector, and foreign patents are pending. The Center for Advanced Technology/Biotechnology at Cornell University supported the construction and maintains a BIBAC Web site in support of BIBAC technology.

NRI Award:

1995; Evalustion and Applicaton of a New BAC Library Vector Designed for Transfer of Large DNA Inserts; 3 years.

Patent:

March 31, 1998; C. M. Hamilton; Binary BAC vector.

Publications:

• Hamilton, C. M., A. Frary, Y. Xu, S. D. Tanksley, and H. B. Zhang. 1999. Construction of tomato genomic DNA libraries in a binary-BAC (BIBAC) vector. The Plant Journal 18(2):223–229.
• Hamilton, C. M. 1998. BIBAC technology: progress and prospects. AgBiotech News and Information 10(1):23N-28N.
• Hamilton, C. M. 1997. A binary-BAC system for plant transformation with high-molecular-weight DNA. Gene 200:107-116.
• Hamilton, C. M., A. Frary, C. Lewis and S. D. Tanksley. 1996. Stable transfer of intact high molecular weight DNA into plant chromosomes. Proceedings of the National Academy of Sciences 93:9975-9979.

Additional Information:

Hamilton was funded in 1995 as an NRI new investigator at Cornell University with additional support from the National Science Foundation and the Department of Energy. She is now employed at Paradigm Genetics, Inc., in Research Triangle Park, North Carolina. The Center for Advanced Technology/Biotechnology maintains a BIBAC Web site at www.bio.cornell.edu/biotechn/BIBAC/BIBAC, that includes how to obtain BIBAC materials, general information, restriction maps, and several detailed protocols. Hamilton replies to all BIBAC correspondence and can be reached at chamilton@paradigmgenetics.com.

Appendix H

Research Needs in Food, Fiber, and Natural Resources

Chapter 5 of the 1989 National Research Council report, *Investing in Research: A Proposal to Strengthen the Agricultural, Food, and Environmental System* (available on the Web at http://books.nap.edu/books/0309041279/html/index.html), provided a detailed list of areas for fundamental research in food, fiber, and natural resources. The committee reviewed the list and found it to be as relevant now as it was 10 years ago. The years have only added to the list of concerns. As part of this study, the committee developed the following list of research needs in food, fiber, and natural resources. Although this list is not as exhaustive and does not provide as much detail as the one in the 1989 report, it is generally consistent with the 1989 conclusions.

PLANTS

Gene and genome interactions and bioinformatics. Mechanisms of interactions of genes and organisms will be identified and will provide the basis of improved growth, metabolism, development, behavior and adaptation. This information will have many applications, from increased plant yield to cleanup of environmental pollution.

Transgenic plants for improved production. Increases in plant yield, disease control, drought tolerance, and many other characteristics will be obtained by constructing transgenic plants. A major research effort will be required to identify new genes with useful characteristics, and to construct and test transgenic plants that carry these genes.

Mechanisms of pest-plant and beneficial-plant interactions. Major advances in the understanding of disease mechanisms and in development of disease-control measures will be stimulated over the next few years by ground-breaking basic research now under way. In addition, there will be increased interest in the direct use of modified beneficial organisms and in identifying the relevant genes for beneficial organisms and transferring them into plants.

Knowledge base for facilitating a new generation of biologically based materials. These would replace such natural products as petroleum, such structural materials as steel, and synthetic textiles. Current trends indicate that it will be possible soon to replace these environmentally sensitive commodities with plant-produced materials that are environmentally safe and renewable. It will be of great consequence to the planet if means can be devised whereby commodities produced by higher plants are coupled with plants that have greater ability to reduce the carbon dioxide load in the atmosphere.

Engineering of plant biosynthetic and metabolic pathways. With the rapidly expanding pool of genes with known functions, it is possible to consider making radical changes in biochemical pathways by introducing new genes and mutating existing genes. It is widely expected that research in this field will yield major returns in development and production of pharmaceuticals and improved plant disease and pest resistance, yield, and other characteristics.

ANIMALS

Gene and genome interactions and bioinformatics. Mechanisms of interactions of genes and organisms will be understood over the next several years and will provide the basis of improved farm-animal growth, metabolism, development, behavior, and adaptation. These findings will permit assessment of future disease potentials in animals and humans and will allow development of diets to avoid them.

Future major progress in the production of livestock species that are important to the US economy will be attainable only if we are able to map animal genomes. This information will provide the basis of regulating various aspects of animal health, growth and development, metabolism, reproduction, and behavior. For example, the ability to identify specific marker genes associated with or predictive of such traits as rate of gain, fecundity, milk production, egg production, and ovulation rate would enable the selection of superior sires and dams in a shorter time than the many years now required through regular genetic approaches. However, mapping the genome of just one species is a time-consuming and costly endeavor. Unless additional funds are made available to the NRI to ensure support for genome mapping, it is unlikely that this aspect of animal research will be carried to completion in a timely manner.

Nutrition research. Devising ways to approach human and animal nutrition in novel ways will become increasingly important. We need research on devising delivery mechanisms and studying their efficacy.

Transgenic and cloned animals. These approaches will greatly increase the efficiency of farm-animal production. It is expected that animals will function as bioprocess reactors to facilitate the introduction of improved nutrients into meat, improved and altered milk composition, and other developments. Research will provide many additional avenues for using the basic advances offered by the ability to clone farm animals and produce transgenic animals.

Animal Reproduction. There is a need to examine the basic physiologic, genetic, and molecular mechanisms that underlie reproductive events that ultimately dictate productivity of economically important livestock and aquatic species. There is also a need to develop improved and new methods of cloning superior livestock.

Reproductive failure continues to be a major cause of revenue loss to livestock producers. Much needs to be learned about the effects of disease and environment on the reproductive system of our livestock species. Much research is needed to identify the underlying causes of embryonic mortality, cystic ovaries, inferior sperm quality, poor conception rate, abortion, and reduced hatchability in poultry. Many of the future research approaches to these problems will of necessity be molecular.

Animal Nutrition. There is a need to examine the potential of genetically engineering rumen microbes that use dietary nutrients effectively for production of meat, milk, and fiber and to research the effects of diet on the evolution and survival of pathogenic rumen and intestinal microorganisms.

It is unlikely that conventional methods of research will yield progress in animal nutrition. Yet, there is the potential for increasing the ability of the animal to use feedstuffs as a source of energy or to use nutrients that now remain undigested. This can be accomplished by using genetically engineered rumen or intestinal microorganisms with specific digestive enzymes. This is a challenging subject that is not being investigated. Land areas now available for grazing or production of forage crops will decline, and livestock might have to be fed foodstuffs that by today's standards are considered to be of poor quality. However, appropriate genetically engineered microorganisms that are deemed nonpathogenic when introduced into livestock species might enable these animals to use poor quality foodstuffs efficiently. Such research will require a concerted effort by several laboratories, will take time, and will be expensive.

Animal–Rangeland Interactions. Western livestock producers are gradually forced out of existence because of various restrictions imposed on the use of riparian areas and western rangelands. Many current environmental issues targeting livestock producers are not based on solid scientific evidence. There is a dire need for research to be conducted to provide well-controlled data on the impact of cattle, sheep, and horses on stream quality under various conditions and on the vegetation and ecology of desert and forested rangelands. There is no concerted effort in the NRI to fund this type of research, which ultimately will affect the economy of the eleven western states.

Animal Health. The US research community must continue to research the organisms that now lay waste to the health of our livestock species. The US should be at the forefront of research on potential emerging diseases, such as BSE, that are or can be transmitted to humans. We need a better understanding of virology as it pertains to infection of livestock by foreign viruses; this is essential not only to develop appropriate vaccines, but also to identify the etiology of viral infections in humans.

Animal Growth and Development. Consumers of meat and meat products are demanding a lean, virtually fatfree product. There is also a growing demand for organically grown animal products, such as meat, milk, and eggs free of hormone residues and antibiotics. The European meat market is closed to American beef producers because of concern about hormones in meat. To provide this type of consumable product, extensive research will be required to define the underlying mechanisms of growth in general and the development of muscle and fat specifically. The identification of specific marker genes that are associated with enhanced meat, milk and egg production can provide an excellent starting point for this research. However, muscle and fat formation involves innumerable complexities which are only tangentially known. This is an area of animal production that can have a serious effect on livestock producers and on the American economy and should be supported by the NRI to a greater extent than it is now.

Aquaculture and Mariculture. Aquaculture and mariculture constitute the most rapidly growing sector of animal agriculture; many new species are added each year. A broadly expanded research program in genomics, nutrition, and reproduction of domestic aquatic species is essential to the health and well-being of these newly emerging industries.

Consumable Animal Products. Research is needed to improve and develop methods of processing, packaging, and marketing of animal products for national and international markets.

Immunology. With today's global travel, diseases are exchanged more rapidly than prophylactic drugs can be devised. We need more basic research on the immunology of humans and farm animals, particularly dedicated to treatment of exotic diseases.

Construction of novel microorganisms. By using genes from existing microorganisms and synthetic genes, new and redesigned microorganisms can be envisioned in the near future. Such altered organisms have the potential for major impact in agricultural settings, for example, rumen microorganisms in cattle.

Gene-based pharmaceuticals and gene therapy. Many experimental approaches, including the use of antisense RNAs and the transient expression of introduced genes, will greatly change how disease and nutritional problems are approached. A great deal of research will be required to deliver such approaches to the marketplace and to undercover new rationales for exploitation.

Evolution of biologic systems. The several genome projects under way or expected will yield unprecedented knowledge applicable to understanding the evolutionary history of humans, animals, plants, and microorganisms. Much of this information will in turn have a major impact on agricultural processes.

NUTRITION, FOOD SAFETY, AND HEALTH

Research on nutrient-drug interactions. People are living longer and using alternative foods and drugs to improve quality or life. We need better information on the effects of excessive nutrient and drug use.

Impact on consumers of phytochemical substances promoted as nutraceuticals. There is a need to isolate and characterize at the molecular level the active agents in traditional and alternative crops and to assess their effects on specific targeted physiologic responses and side effects.

New and resurgent pathogens in foods. Greater numbers of microbial food-contamination problems are arising. We need to know more about the microorganisms in question—their biology and mechanisms for their control. Information is needed on the source of pathogens encountered in production, harvesting, processing, and distribution of plant, animal, and marine products.

Pasteurizing and sterilizing of foods. New methods need be developed in food processing with regard to advanced sterilization techniques. Opportunities for the wider use of high-pressure preservation and pulsed electric-power discharges in food preservation need to be assessed. Molecular modifications of and effects on quality and hazardous microorganisms need to be carefully assessed.

Identification and modification of allergens in foods. Rapid, simple, and cost-effective tests for the presence of known and unidentified allergens in foods should be developed. It is also necessary to evaluate mechanisms of allergenicity and develop methods of processing to neutralize allergen hazards in foods.

Probiotic development. With more antibiotics becoming obsolete, we need to develop more natural antibiotics to maintain homeostasis.

NATURAL RESOURCES AND THE ENVIRONMENT

Water quality. As increasing numbers of water supplies become unacceptable for human consumption or food processing, the need for research to characterize contaminants and facilitate their removal is clear.

Animal-waste handling. Concentration of animal production has led to major problems with respect to odor emanation and the handling of waste products. Major research efforts are warranted.

Environmental impacts. Great developments are occurring in devising methods to ameliorate toxic-waste deposits and other environmental insults. The success of American agriculture in coming decades will depend on solution of current toxic problems and seeing that agricultural practices do not produce new environmental disasters.

Impact of environmental and biotechnologic modifications of plants and animals on microbial ecology of food products. There will be an increasing need to evaluate the safety and quality of foods that come from genetically modified organisms in contrast with traditional sources.

Bioprocess engineering. Modifications of plants, animals, and microorganisms in the next century can be expected to have major ramifications for the production of commodities and for ecologic concerns, such as reducing waste production. We predict that major advances can be made in bioengineering of the organisms that we deal with in agricultural settings

Biodiversity. Agriculture is often criticized as being incompatible with biodiversity. Yet basic research in conservation biology suggests that landscape diversity can be as important as species diversity in plant communities for maintaining diverse populations of insects and vertebrates in a geographic region. In fact, agricultural communities can benefit directly from landscape diversity in that this diversity often includes beneficial organisms that prey on crop pests, pollinate crops, and provide other ecosystem services for cropped fields that have economic benefit. The design of landscapes—for example, development and tests of theory that predicts the optimal proportion of native to cropped habitat or the best positions of such habitat within landscapes—is a basic-research question that is not now being adequately addressed.

Soil biodiversity is another major new branch of research that requires additional support. We have only recently acquired the molecular tools for gauging the complexity of the soil biologic community and have learned that only 3%–5% of soil microbial taxa have been identified and described. What constitutes the unknown 95%? Does it differ among ecosystems and management practices? Does it have functional significance for the forest, rangeland, or cropping system that supports it? Those are basic questions with enormous potential impact on the management and protection of agricultural resources.

Understanding biodiversity also requires an understanding of basic population genetics, and many fundamental questions in population genetics that are relevant to agriculture are being insufficiently addressed. How will gene flow in native populations retard or accelerate the movement of genes from genetically modified organisms—for example, from Bt corn or glyphosate-resistant soybeans—into native plant populations? Will it matter that specific genes escape? Will they persist? How long will it take native pest populations to develop resistance to engineered traits in the genetically modified organisms? Answers to those basic ecologic and evolutionary questions will help to define the efficacy and safety of genetically modified crops. Answers to them will also help to determine strategies for protecting native and cropped communities from colonization by exotic organisms in general—an increasingly important threat to crop and forest productivity in this age of global trade.

Weather and climate interactions in agricultural systems. Basic research in coming decades will uncover additional approaches to minimizing agricultural losses during weather disasters. Genes for increased cold tolerance in plants are already affecting losses from cataclysmic freezes, and many other examples can be predicted.

Global change and agriculture. Agriculture both affects and is affected by many of the environmental changes that fall under the global-change rubric. Many of the practical issues that have emerged over the last decade are being addressed with mission-oriented funding from a number of agencies, including the US Department of Agriculture (USDA). These issues range from evaluating the impact of continent-scale transport of atmospheric contaminants and gauging the effects of increased climate variability and changing hydrologic cycles on crop productivity to evaluating the potential for agricultural soils to sequester atmospheric CO_2. Solutions of those and the scores of other important global environmental-change problems depend implicitly on a thorough understanding of the principles that govern the patterns and processes affected by change. Such understanding is provided by basic research in such topics as soil organic matter, environmental plant physiology, and environmental modeling.

Soil Organic Matter Dynamics. Cropped soils typically lose 40%–60% of their carbon after 40–60 years of cultivation. The recovery of the carbon and the potential storage of additional carbon has been widely touted as a potential mitigator of atmospheric CO_2 buildup. But fundamental mechanisms of carbon storage in soils—for example, the relative importance of physical protection by clays and microaggregates versus chemical protection by humic and other organic substances and the contribution of recalcitrant versus active fraction to total carbon stores in different ecosystems—are poorly understood. Only since the 1993 creation of the small NRI Soils Program has basic, peer-reviewed soil-carbon research had an important source of potential support in USDA. Basic soils research is in general poorly supported at the national level; soil carbon is one of many competing needs.

Environmental Plant Physiology. Responses of plants to changes in atmospheric and soil chemistry are key determinants to the effects of global change on ecosystems. Many responses are interactive and require an ecosystem context in which to understand them sufficiently to suggest management solutions. But many basic ecosystem interactions are too poorly understood to gauge the effects of change. For example, increased CO_2 in the atmosphere leads to changes in leaf chemistry in many tree species; do these changes affect insect herbivory or leaf-litter decomposition rates? If so, how will the changes affect other trophic levels and soil nutrient availability, and eventually plant susceptibility to insect outbreaks, fire, and drought? Is the response of forests to nitrogen saturation ameliorated by increased CO_2 ? Is the response of annual plants—both crops and weeds—to increased CO_2 fundamentally different from that of woody perennial plants at either the plant or the ecosystem level? Answers to those questions require fundamental knowledge that is being gained very slowly via poorly funded basic-research programs in agricultural ecosystems.

Environmental Modeling. The complexity of ecosystems and differences in their responses to climatic variability suggest that process-based, quantitative models will eventually be the best way to predict the effects of human activities on ecosystem structure and function and to suggest the likely effects of different management scenarios and therefore best-management solutions. However, basic research into quantitative modeling is funded as a small part of the NRI Agricultural Systems program. To effectively link existing crop and forest models to models of soil biogeochemistry, hydrologic transport, and atmospheric chemistry—and then link these models to economic, land-use, and other social models—will require substantive basic research not now budgeted for.

Nitrogen-Use Efficiency. The efficient use of nitrogen in cropping systems is essential for protecting downstream ecosystems from environmental harm while maintaining high agricultural productivity in cropped fields. Nitrogen limits crop growth, but only 50%–60% of nitrogen applied to crops is taken up by them. Most of the remainder is lost to groundwater and surface water as nitrate or to the atmosphere as dinitrogen or the greenhouse gas nitrous oxide. The environmental and economic consequences can be large: an increasing proportion of drinking water, especially in rural areas, exceeds Environmental Protection Agency health standards for nitrate contamination, and nitrate in surface water eventually makes its way to coastal areas, where it has been linked to unwanted effects—hypoxic zones that can depress such populations as shrimp in the Gulf of Mexico, outbreaks of such toxic algae as those in Chesapeake Bay, and so on.

Improving nitrogen-use efficiency on the field scale has been an elusive goal; efficiency has changed little since it was first measured in the 1950s. Changes in tillage practices, the application of site-specific farming methods, the introduction of nutrient catch crops in the cropped field or of riparian vegetation downstream, and increases in the efficiency of plant nitrogen uptake in the rhizosphere all depend on a better fundamental understanding of the ecologic interaction among crops, soil organisms, and the set of physical and chemical conditions that define the plant-soil environment. Opportunities for peer-reviewed, competitive funding of basic, integrated research in crop ecosystems are largely lacking outside the small NRI programs in Natural Resources and Environment.

Wildlife in agricultural systems. The conflict of expanded agricultural productivity and the desire for environmental preservation, including that of wildlife, necessitates further research to devise new methods and approaches.

Space. Agriculture and agricultural research will play a major part in space exploration because of obvious needs for food, fiber, and waste disposal. Engineered microbial lines and bioprocessing will have large contributions to make and need to be studied now in the context of space flight.

ENHANCING VALUE AND USE OF AGRICULTURAL AND FOREST PRODUCTS

Development of microtechnology for separation and analysis of biologic molecules using microfabrication and nanotechnology. These fields are progressing extremely rapidly and promise to affect approaches to use of agricultural and forest products in new ways.

Impacts of organic farming. The safety and quality of organically produced products need to be compared with those of conventionally produced commodities and characterized at the molecular level.

Bioprocess engineering for agricultural products. Integrated research is needed to combine molecular-biology techniques for tailoring plants to the generation of specific value-added products with postharvest processing steps that will enable cost-effective recovery of the products in appropriately located and sized bioprocessing plants.

Metabolic pathway analysis. Structure-function studies of proteins that are generated by genes discovered in the Plant Genome Project should be undertaken.

MARKETS, TRADE, AND RURAL DEVELOPMENT

Knowledge base to prepare for biologic terrorism or deliberate attempts to degrade the biosphere or agroecosystems. There is a great need to expand our knowledge base—for instance, on pathogenic microorganisms—to forestall attempts at biologic terrorism.

Globalization of the economy. Greater broadening of economies has major effects on US agriculture. There is a need for increased research to explore the bases and ramifications of this increasing trend.

Economic and social consequences of environmental regulation. Studies should assess the benefits and costs of government regulations that affect agricultural production and the environment, design and evaluate alternative policies and institutions to mitigate negative environmental impacts of production agriculture, and develop more quantitative and qualitative tools for assessing nonmarket goods.

Risk-management tools and financial management. Studies should assess ways to measure and manage risk in a new, globalized, vertically coordinated food system; analyze specific risk-management strategies, instruments, and portfolios; and assist farmers and lenders in adopting improved financial accounting and reporting systems.

Impacts of the changing farm and agribusiness structure. Studies should analyze the forces driving structural change and concentration and their effects on the economic performance of vertically coordinated farming and agribusiness; determine the effects of vertical coordination on market access, bargaining power, concentration, location of production, financial arrangements, rural communities, and the environment; and analyze the relationship between value-added agricultural commodities and new-product development, producer profitability, risk, and market access.

Evaluate trade policies and barriers. Studies should assess the benefits, costs and other implications of trade policies, government regulation, and institutional barriers to international trade; evaluate the relationships among trade, natural resources, and the environment; and enhance understanding of the economic impacts and consequences of trade.

Economic and rural community development programs. Studies should create improved information to assist local governments in cost-effectively meeting demands for public services, financing public programs, providing infrastructure needs, and designing incentives for private-sector initiatives and involvement; improve understanding of the roles of human capital, social capital, and life-long learning in rural economic development; and ascertain the impacts of government programs on rural poverty.

Effects of changes in consumer demand on health, nutrition, and food safety. Studies should assess the benefits and costs of public policies and government regulations that affect health, nutrition, and food safety; assess consumer preferences and demands and their implications for production and marketing practices in the food system; and increase multidisciplinary analysis of food-science issues.

Economic and social impacts of consolidating research and extension programs. Studies should assess opportunities for regionalization of research and extension programs, change the reward systems for agricultural research to value multidisciplinary and applied work more highly, and achieve greater coordination among research and extension, including involvement by stakeholders in priority-setting, planning, and program evaluation.

Information technologies and communication systems. Studies should ascertain the benefits and costs of public versus private information and the implications for delivery systems for agricultural research results and education, redesign the delivery systems of the Cooperative Extension Service for more effective and timely performance, and evaluate the value and use of precision technology and information in agricultural production.

Economic and social impacts of biotechnology: Studies should analyze how biotechnology affects farm size, production efficiency, competitiveness, trade potential, and other elements of economic performance in agriculture; evaluate the public- versus private-sector roles in the development of biotechnology; and enhance the public's understanding of the benefits and risks associated with biotechnology.

Development of human capital. Studies should place greater emphasis in undergraduate curricula and public education on understanding the global economy, renew the emphasis on competitiveness as a key economic concept in agriculture and agribusiness curricula, use more "real-life" and experiential learning in the classroom.

Appendix I

Grant Performance Follow-on for Quality Evaluation

(Examples of what could be collected)

Project Title
Date of initiation of the NRI grant
Additional support in conjunction with or as a result of this grant

- Agency (for example, Department of Energy, National Institutes of Health, National Science Foundation, State Agricultural Experiment Stations)
- Private Sector
- Other

Presentations resulting from research funded by NRI grant

- Referred journal citations
- Other articles
- Theses
- Citation Index/Citation Record

News releases on findings of research funded by grant
Disclosures for patents

- Patents pending
- Patents resulting from NRI grant
- Licenses

Interactions with other scientists in private sector, universities, government
Testimonials
Transfer of findings to possible application

References

Alston, J.M., and P.G. Pardey. 1996. Making Science Pay: The Economics of Agricultural R&D Policy. Washington, D.C: AEI Press.

Baldwin, W., and P. McCardle. 1996. Peer review at NIH in the 21st century. FASEB Journal 10(December 1996):1563–1568.

Barry, P.J. 1997. Rates of return to public investment in agricultural research and education. CHOICES Fourth Quarter.

CED (Committee for Economic Development). 1998. America's Basic Research: Prosperity Through Discovery.

Chubin, D. 1994. Grants peer review in theory and practice. Evaluation Review 18(Vol. 1, February):20–30.

Chubin, D. 1998. Looking Backward, Then Forward at American Research Universities. Pp. 375–384 in A.H. Teich et al., eds., AAAS Science and Technology Policy Yearbook. Washington DC: American Association for the Advancement of Science

Chubin, D., and E.J. Hackett. 1990. Peerless Science: Peer Review and U.S. Science Policy. Albany: State University of New York Press.

Daily, G., P. Dasgupta, B. Bolin P. Crosson, J. du Gerny, P. Ehrlich, C. Folke, A.M. Jansson, B. O. Jansson, N. Kautsky, A. Kinzig, S. Levin, K.-G. Maler, P. Pinstrup-Andersen, D. Siniscalco, B. Walker. 1998. Food production, population growth, and the environment. Science 281:1291–1292.

Fuglie, K., N. Ballenger, K. Day, C. Klotz, M. Ollinger, J. Reilly, U. Vasavada, and J. Yee. 1995. Agricultural research and development: public and private investments under alternative markets and institutions. Agricultural Economic Report No. 735. Economic Research Service, US Department of Agriculture.

Kelman, A., and R.J. Cook, 1996. The role of a competitive research grants program for agriculture, food, and natural resources. Bio Science 46:533–540.
Klotz, C., K. Fuglie, and C. Pray. 1995. Private-sector agricultural research expenditures in the United States, 1960–92. Agricultural Economics Report No. 9525, Economic Research Service, US Department of Agriculture.
Kostoff, R. 1996. The Principles and Practices of Peer Review. Science and Engineering Ethics 3:19–34.
Kostoff, R. 1997a. Peer Review: The Appropriate GPRA Metric for Research. Science 277:651–652.
Kostoff, R. 1997b. Research Program Peer Review: Principles, Practices, Protocols. (http://www.dtic.mil/dtic/kostoff/index.html).
Lipton, K.L., W. Edmondson, A. Manchester. 1998. The food and fiber system: Contributing to the U.S. and world economies. Economic Research Service, US Department of Agriculture, Agricultural Information Bulletin No. 742. Washington, DC, July 1998. Available online: <http://www.econ.ag.gov/epubs/pdf/aib742/index.htm>
MacIlwain, C. 1998. Plant scientists want focus on the best and brightest. Nature 394:210–211.
May, R. 1998. The scientific investments of nations. Science 281:49–51.
McGeary, M., and P.M. Smith. 1996. The R&D Portfolio: A Concept for Allocating Science and Technology Funds. Science 274:1484–1485.
Metzger, N., and R.N. Zare. 1999. Interdisciplinary Research: From Belief to Reality. Science 283:1096–1097.
Mowery, D.C., N. Rosenberg. 1989. Technology and the Economic Pursuit of Growth. Cambridge: Cambridge Press.
National Science Foundation. 1999. Survey of Federal Funds For Research And Development: Fiscal Years 1996, 1997, 1998. Washington DC: US Government Printing Office.
Nature. 1998. Genetic Revolution Overdue at the USDA [editorial]. Nature 394:207.
NRC (National Research Council). 1972. Report of the Committee on Research Advisory to the U.S. Department of Agriculture. Washington, DC: National Academy Press.
National Research Council. 1989. Investing in Research: A Proposal to Strengthen the Agricultural, Food and Environmental System. National Academy Press: Washington, DC.
National Research Council. 1994. Investing in the National Research Initiative: An Update of the Competitive Grants Program of the U.S. Department of Agriculture. National Academy Press: Washington, DC.
National Research Council. 1995. Allocating Federal Funds for Science and Technology. National Academy Press: Washington, DC.
National Research Council. 1998. Assessing the Value of Research in the Chemical Sciences. National Academy Press: Washington, DC.
National Research Council. 1999. Evaluating Federal Research Programs. National Academy Press: Washington, DC.
Public Law 101-624, 101st Congress, Federal Register.
Science. 1998a. The Biocomplex World of Rita Colwell. 281:1944–1947.
Science. 1998b. 1999 Budget: One Step Forward, Two Back. 282:392–393.
Science. 1999. 2000 Budget: Thanks to NIH, R&D Ends Up With 5% Boost. 286:1836.
US Congress, Office for Technology Assessment. 1977. Organizing and Financing Basic Research to Increase Food Production. Washington, DC: US Government Printing Office.

US Congress, Office for Technology Assessment. 1991. Federally Funded Research: Decisions for a Decade. Washington, DC: US Government Printing Office.

US Congress, Office for Technology Assessment. 1995. Challenges for U.S. Agricultural Research Policy. Washington, DC: US Government Printing Office.

US Department of Agriculture. 1997. National Research Initiative Competitive Grants Program Application Kit. July 1997. Washington, DC: US Department of Agriculture.

US Department of Agriculture, National Research Initiative Competitive Grants Program, personal communication, June 26, 1998.

US Department of Agriculture, National Research Initiative Competitive Grants Program, personal communication, September 1, 1998.

US Department of Agriculture, National Research Initiative Competitive Grants Program, personal communication, March 23, 1999.

US General Accounting Office. 1996. Agricultural Research: Information on Research System and USDA's Priority Setting, GAO/RCED 96-92. P. 44 and App. V. Washington, DC: US Government Printing Office.

US General Accounting Office. 1999. Peer review practices at federal science agencies vary. GAO/RCED 99-99. Washington, DC: US Government Printing Office.

US House of Representatives, Committee on Science. 1998. Unlocking Our Future: Toward a New National Science Policy [online]. Available online: <http://www.house.gov/science_policy_report.htm>. [September 24].

About the Authors

Thomas N. Urban is the former chairman and CEO of Pioneer Hi-Bred International, Inc., and currently chairman of the Board of Trustees, Carnegie Institution of Washington. Mr. Urban joined Pioneer in 1960 and became president in 1979 and chairman, president, and CEO in 1984. In 1992, he was named Iowa Business Leader of the Year. On leaving Pioneer in 1995, Mr. Urban became a visiting professor at the Graduate School of Business Administration, Harvard University. He has served on numerous advisory committees, including the World Agriculture Development Foundation, President-elect Clinton's transition team, and the Council on Foreign Relations. Mr. Urban has also served on National Academies groups, including the Working Group on New Alliances and Partnerships: Enhancing the Utilization of Scientific and Engineering Advances and the Government-University-Industry Roundtable. He is a member of the National Research Council's Board on Agriculture and Natural Resources. Mr. Urban received his MBA degree from the Harvard Graduate School of Business and his BA from Harvard University.

Peter J. Barry is professor of agricultural finance and director of the Center for Farm and Rural Business Finance at the University of Illinois, Urbana. He is a past president and a fellow of the American Agricultural Economics

Association. Dr. Barry is also the past chair of the ESCOP Social Sciences Subcommittee and the current chair of the Council on Food, Agricultural and Resource Economics (C-FARE). He has been a professor at Texas A&M University and the University of Guelph; editor of the *American Journal of Agricultural Economics*; and consulting editor for the *European Review of Agricultural Economics*. Dr. Barry received his PhD, MS, and BS degrees in agricultural economics from the University of Illinois.

Francis F. Busta is professor and former head of the Department of Food Science and Nutrition at the University of Minnesota, St. Paul. He is a consultant to 3M, the International Consortium of Food Quality Professionals, Inc., and the Proctor and Gamble Company. He is a long-time member of the International Commission on Microbiological Specifications for Food. His research interests include environmental stress on microorganisms, influence of food systems on growth and survival of microorganisms, thermal processing, general microbiologic aspects of food processing and preservation, probiotic control of sporeformers and cancer, and hazard-analysis critical control point approaches to quality management. Dr. Busta studied food science at the University of Illinois, where he received his PhD degree; and dairy industries and bacteriology at the University of Minnesota, where he received his MS and BA degrees.

Mary-Dell Chilton is senior staff scientist at Novartis Seeds in Research Triangle Park, North Carolina. She earned her PhD in chemistry on the Urbana campus of the University of Illinois in 1967. After postdoctoral studies and various research appointments at the University of Washington, she joined the faculty of Washington University, in St. Louis, in 1979. Her research group in collaboration with others produced the first transgenic tobacco plants in 1982. She joined Ciba-Geigy in 1983 as executive director of agricultural biotechnology. She held administrative positions at Ciba for 11 years, and she has returned to active engagement in bench research at Ciba, now Novartis through merger. Her honors include the Rank Prize in Nutrition (1986), election to the National Academy of Sciences (1985), the David Gotlieb Medal (1986), the Hendricks Medal (1987), and election to the American Academy of arts and Sciences (1993). She was awarded the degree of Doctor Honoris Causa by the University of Louvain in 1983. She is a fellow of the American Academy of Microbiology (1994) and has served on the North Carolina Board of Science and Technology since 1986. Her current research is directed to improving the technology for introducing new genes into crop plants.

Daryl E. Chubin is senior policy officer in the National Science Board Office of the National Science Foundation (NSF) in Arlington, Virginia. While at NSF, he has also served as director of the Research, Evaluation, and Communication Division. In 1997, he was on detail as assistant director for social and behavioral sciences and education at the White House Office of Science and Technology Policy. Dr. Chubin's professional interests have focused on the social and political dimensions of science and technology as related to education, human resources, interdisciplinary teamwork, mass media, and public policy. He has taught at five universities, served on the staff of the congressional Office of Technology Assessment, is a founding member of the Society for Social Studies of Science, and is president of the Commission on Professionals in Science and Technology and chair-elect of the American Association for the Advancement of Science (AAAS) Section on Societal Impacts of Science and Technology. Dr. Chubin received his PhD in sociology from Loyola University in Chicago.

Robert J. Collier is professor in the Department of Animal Sciences at the University of Arizona in Tucson. He was previously senior fellow and dairy research director at Monsanto Company in St. Louis, Missouri. His responsibilities included directing all preclinical and clinical dairy research required for commercial approval of bovine somatrotropin in the United States and Canada for the Animal Sciences Division of Monsanto and directing the discovery research program for novel products for the dairy industry. His research interests focus on nutritional physiology, the physiology of lactation, and the effects of environment on dairy cattle and their endocrine systems. Dr. Collier completed an National Institutes of Health postdoctoral fellowship at Michigan State University. He received his PhD in physiology from the University of Illinois and his MS and BS degrees from Eastern Illinois University.

Noel T. Keen is distinguished professor in the Department of Plant Pathology at the University of California, Riverside (UCR). He is also director of the UCR Biotechnology Center. Dr. Keen's research has been key in understanding the molecular and biochemical basis of virulence, pathogenicity, resistance, and gene-for-gene relationships in plant-pathogen interactions. His team's research pioneered the isolation and characterization of a race-specific elicitor of resistance encoded by a single avirulence gene. Dr. Keen was elected to the National Academy of Sciences in 1997. He serves as an editor for *Plant Physiology* and the *Journal of Bacteriology*. Dr. Keen studied plant pathology at the University of Wisconsin, where he received his PhD degree. He received his MS and BS degrees from Iowa State University.

Michael R. Ladisch is professor in the Agricultural and Biological Engineering Department and Director of the Laboratory of Renewable Resources Engineering at Purdue University. His research addresses fundamental topics in separations and reaction kinetics with current projects in bioseparations, kinetics of biochemical reactions, chemical reaction engineering, and biomass conversion. Dr. Ladisch has served as a member of US delegations and advisory panels to review biotechnology programs in Russia, Thailand, China, and Japan. He chaired the National Research Council Committee on Bioprocess Engineering. He is a former chair and long-range program coordinator of the Biotechnology Division of the American Chemical Society. Dr. Ladisch received his PhD and MS degrees in chemical engineering from Purdue University and his BS degree from Drexel University. Dr. Ladisch was elected to the National Academy of Engineering in 1999.

G. Philip Robertson is professor in the Department of Crop and Soil Sciences and the W.K. Kellogg Biological Station at Michigan State University, Hickory Corners. Dr. Robertson's primary research interests include nitrogen availability in managed and natural ecosystems, particularly agricultural systems; trace-gas fluxes from agricultural landscapes, in particular nitrous oxide and methane; spatial variability of soil resources and its relationship to community structure and productivity; and the functional significance of soil microbial diversity. He directs the National Science Foundation Long-Term Ecological Research site in row-crop agriculture and recently served as director of the US Department of Agriculture Fund for Rural America Environment Program. Dr. Robertson received his PhD in ecology and evolutionary biology from Indiana University and his BA from Hampshire College.

Ronald R. Sederoff is distinguished professor and Edwin F. Conger Professor of Forestry at North Carolina State University. He is also an associate member of the Department of Genetics and the Department of Biochemistry and director of the Forest Biotechnology Group at the university. Dr. Sederoff's research interests focus on the molecular genetics of forest trees to accelerate breeding. His research group was the first to transfer a gene into a conifer. They have also developed methods for genomic mapping of individual trees, applied those methods to complex-trait analysis, and used developmental specialization in wood formation to investigate the biochemical and genetic basis of cell wall biosynthesis. Dr. Sederoff received his PhD, MA, and BA degrees in zoology from the University of California, Los Angeles. He completed postdoctoral studies at the Institute de Biologie Moleculaire and a sabbatical in forest genetics with the US Forest Service at Berkeley. Dr. Sederoff was elected to the National Academy of Sciences in 1995.

William W. Simpkins is associate professor of geology in the Department of Geological and Atmospheric Sciences at Iowa State University, Ames. Dr. Simpkins's research interests are in the hydrogeology of till, agricultural water quality, hydrogeology of riparian zones and buffer strips, and isotope hydrology. He is the associate editor of *Ground Water*. Dr. Simpkins has been a research geologist with the Bureau of Economic Geology of the University of Texas at Austin where he studied the geomorphology and hydrology of high-level nuclear-waste disposal in salt in the Texas Panhandle. He also worked as research liaison between the Bureau of Economic Geology and the US Department of Energy Salt Respository Project Office in Columbus, Ohio. Dr. Simpkins's educational background includes studies at the University of Wisconsin, where he received his PhD degree in geology and geophysics and two MS degrees—one in geology and geophysics, the other in water-resources management. He received his BA degree in geology from Augustana College in Illinois.

Robert E. Smith is president of R E Smith Consulting, Inc., in Newport, Vermont. He has been president of the Institute of Food Technologists, senior vice president for corporate research at Nabisco, Inc., and senior vice president for research and development at Del Monte Corporation. His research focuses on the nutritional quality of human and pet foods, particularly protein and amino acid requirements and interrelationships. Dr. Smith received his PhD degree in animal science, with an emphasis on nutrition and biochemistry, from the University of Illinois. He received his MS and BS degrees in animal and poultry nutrition from McGill University.

Fredrick Stormshak is distinguished professor at Oregon State University. He holds joint appointments in the Department of Biochemistry/Biophysics and the College of Veterinary Medicine. He has been interim head of the Department of Animal Science and associate director of the Oregon Agricultural Experiment Station. Dr. Stormshak's research interests focus on various aspects of uterine and ovarian function in domestic animals. His laboratory contributed initial data demonstrating that ovarian hormones in cattle and sheep act locally to regulate uterine vascular function. He was one of the first researchers to quantify uterine estrogen receptors in sheep and mink and to conduct research on hormonal regulation of uterine steroid-receptor synthesis in these species. Dr. Stormshak is past president of the Society for the Study of Reproduction. He earned his PhD degree from the University of Wisconsin at Madison and his MS and BS degrees in dairy science from Washington State University.

Anne K. Vidaver is professor and head of the Department of Plant Pathology at the University of Nebraska, Lincoln. She also serves as director of the Center for Biotechnology at the university. Dr. Vidaver's research interests focus on biology, genetics, and control of phytopathogenic bacteria; applications of endophytic bacteria; and research-policy issues, especially in the biologic sciences. Dr. Vidaver is associate editor of the *World Journal of Microbiology and Biotechnology* and special reviewer for *Plant Disease* and *Phytopathology*. She is a past member of the National Research Council Board on Agriculture and Natural Resources and of the US Department of Agriculture National Agricultural Research, Extension, Education, and Economics Advisory Board. Dr. Vidaver studied bacteriology at Indiana University, where she received her PhD and MS degrees. She received her BA degree in biology from Russell Sage College.

RECENT PUBLICATIONS OF THE BOARD ON AGRICULTURE AND NATURAL RESOURCES

Policy and Resources

Genetically Modified Pest-Protected Plants: Science and Regulation (2000)
The Use of Drugs in Food Animals: Benefits and Risks (1999)
Ensuring Safe Food: From Production to Consumption (1998)
Agriculture's Role in K-12 Education (1998)
Designing an Agricultural Genome Program (1998)
Brucellosis in the Greater Yellowstone Area (1997)
Forested Landscapes in Perspective: Prospects and Opportunities for Sustainable Management of America's Nonfederal Forests (1997)
Precision Agriculture in the 21st Century: Geospatial and Information Technologies in Crop Management (1997)
Wood in Our Future: The Role of Life-Cycle Analysis (1997)
Colleges of Agriculture at the Land Grant Universities: Public Service and Public Policy (1996)
Ecologically Based Pest Management: New Solutions for a New Century (1996)
Colleges of Agriculture at the Land Grant Universities: A Profile (1995)
Investing in the National Research Initiative: An Update of the Competitive Grants Program in the U.S. Department of Agriculture (1994)
Rangeland Health: New Methods to Classify, Inventory, and Monitor Rangelands (1994)
Soil and Water Quality: An Agenda for Agriculture (1993)
Managing Global Genetic Resources: Agricultural Crop Issues and Policies (1993)
Site Visit Report of the Scientific Council to the Plant Gene Expression Center, Feb. 24–26, 1993 (1993)
Pesticides in the Diets of Infants and Children (1993)
Managing Global Genetic Resources: Livestock (1993)
Sustainable Agriculture and the Environment in the Humid Tropics (1993)
Agriculture and the Undergraduate: Proceedings (1992)
Water Transfers in the West: Efficiency, Equity, and the Environment (1992)
Managing Global Genetic Resources: Forest Trees (1991)
Managing Global Genetic Resources: The U.S. National Plant Germplasm System (1991)
Sustainable Agriculture Research and Education in the Field: A Proceedings (1991)
Site Visit Report, Scientific Council to the Plant Gene Expression Center, August 30–31, 1991 (1992)
Toward Sustainability: A Plan for Collaborative Research on Agriculture and Natural Resource Management (1991)
Wild Horse Populations: Field Studies in Genetics and Fertility (1991)

Site Visit Report, Scientific Council to the Plant Gene Expression Center, January 8–9, 1990 (1990)
Technology and Agricultural Policy: Proceedings of a Symposium (1990)

Publications of the Committee on Animal Nutrition, 1990–1998

Nutrient Requirements of Swine, Tenth Revised Edition (1998)
The Role of Chromium in Animal Nutrition (1997)
Nutrient Requirements of Beef Cattle, Seventh Rev. Ed. (1996)
Building a North American Feed Information System (1995)
Nutrient Requirements of Laboratory Animals, Fourth Rev. Ed. (1995)
Metabolic Modifiers: Effects on the Nutrient Requirements of Food-Producing Animals (1994)
Nutrient Requirements of Poultry, Ninth Rev. Ed. (1994)
Nutrient Requirements of Fish (1993)